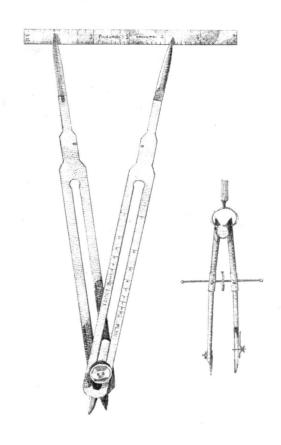

The Wonderful World of Mathematics

CARL A. RUDISILL LIBRARY
LENOIR RHYNE COLLEGE

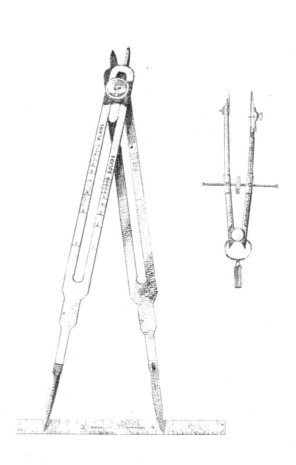

CONTENTS

The Beginning: **Time and Tally** 6

Ancient Egypt: **Taxes and Triangles** 12

Babylon and Assyria: **Square and Circle** 20

Phoenician Voyages: **Stars and Steering** 26

Greece and Rome: **Proof and Progress** 30

The Moslem Empire: **Numbers and Nothing** 44

Western Europe: **Graphs and Gravity** 52

The Industrial World: **Power and Precision** 62

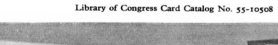

Library of Congress Card Catalog No. 55-10508

Produced by Rathbone Books, London — Printed in Great Britain by L. T. A. Robinson, Ltd., London

THE WONDERFUL WORLD OF MATHEMATICS

LANCELOT HOGBEN

Art by

André
Charles Keeping
Kenneth Symonds

maps by Marjorie Saynor

GARDEN CITY BOOKS · GARDEN CITY · NEW YORK

FIRST PUBLISHED IN THE UNITED STATES OF AMERICA IN 1955

Time and Tally

FOR MY one deer you must give me three of your spearheads. The earliest men and women like ourselves lived about twenty-five thousand years ago. They could say all this with their hands, simply by pointing one finger at the deer and three at the spearheads. The primitive way of counting with one finger for one thing and three fingers for three things, was the only kind of arithmetic they knew. For thousands of years such people thought of any quantity greater than three as a heap or pile.

They had no towns, no villages. They were wanderers who trekked from place to place in search of animals and birds to hunt and of berries, roots, and grain to gather. The only goods they possessed were the skins of animals, to protect them from the cold night air, a few hunting weapons, crude vessels to hold water, and perhaps some kind of lucky charm, such as a necklace of bear's teeth or sea-shells.

There was no need for them to know much arithmetic. Even their simple finger-counting was useful only on the rare occasions when they wanted to exchange goods with the members of some other tribe.

Every night, at places as far north as Great Britain you may see a star-cluster circle slowly round the Pole Star which scarcely moves.

Much more important to these hunters and food-gatherers was a knowledge of the seasons and of direction. Knowledge of the seasons could help them to forecast when the nuts and berries were beginning to ripen in some far-off forest, and a knowledge of direction would help them to find their way there. With neither calendars nor maps to help them they had to learn these things slowly, by long experience through trial and error.

While they were wandering through countryside they knew, they could find their way by remembering the positions of familiar hills, lakes and streams; but when drought or hunger drove them to seek new hunting grounds they had only the sun, moon and stars to guide them.

Tribes living near the sea might notice that the sun seemed to rise each morning out of the waves and set each night behind some distant line of hills. They could find their way to the sea by marching towards the rising sun, or to the hills by marching towards the setting sun. But this bit of knowledge would give them only a very rough-and-ready guide, for the sun's rising and setting positions change from season to season.

The stars of the night sky offered a much more reliable clue to direction but it must have taken many, many years for the wise men or women of these early tribes to discover it. We can imagine them, after the day's hunting was over, sitting by the opening of a shelter or the mouth of a cave and gazing up into the starlit sky. After a time they would notice certain clusters of stars that formed simple patterns which they could pick out night after night. These star-clusters seemed to trace part of a circular path across the sky, moving slowly round like the hands of some giant clock.

Some clusters seem to circle around a fixed point in the northern sky. There lies what we now call the North Pole Star, which scarcely changes its position in the night sky in a hundred years. Since it seems to be fixed, it is a kind of signpost. Nightlong this star shows us where what we call north lies, if we can spot it among all the hundreds of other stars that shine and twinkle in the sky.

Like us, the hunter of twenty-five thousand years ago could locate this signpost by spotting a cluster of seven stars, shaped rather like a big dipper or an ancient plough. This cluster circles around the Pole Star. Wherever we see it in the night sky, two of its stars point almost directly to the Pole Star. If we go the way they point, we are going northwards.

By night a man can rely on the Pole Star to guide him north.

From its full round, the moon changes a little every night, growing slimmer until it disappears, then gradually back to the full.

Sun, moon and stars were not only man's first signposts, they were also his first clock. During the day, the early hunter living north of the tropics would see the long morning shadows point westward. He would watch them grow gradually shorter until the sun reached its highest

The first calendar: notches cut to record changes of the moon.

point in the heavens at noon. As the sun sank lower, he would then see the shadows, now pointing eastward, slowly lengthen again. By noticing the length of shadow he could roughly tell what we now call the time of day.

Watching by the camp fire, these early folk would notice that the moon when full is highest in the sky just halfway through the night. In time, the more observant ones would also learn to judge the night hours by following the course of certain star-clusters which circle around the Pole Star.

To measure longer periods of time, our first forefathers must have relied on the moon. Night by night they saw how it gradually changes from a full disc of silver to a slim crescent and then disappears altogether. After a few dark nights, it reappears as a crescent and slowly grows again to its full size.

Just as the full moon was rising, a hungry tribe might pitch its tents near a wood whose boughs were laden with sour, green berries. The wise ones might say: Let us not touch these berries now; let us come back when the moon is once more full; then they will be black and good for plucking. The clan would then wander far afield in search of other food. Somehow they had to make sure of getting back at the right time. To do that, they would need to count the days.

Time flies, and counting days or months is not like counting dead deer or bear's teeth. We cannot make days stand in a row while we count them on our fingers. Our forefathers most likely first solved the problem by cutting a notch on a tree, a stick or a stone to mark the passage of each day: one notch – one day, two notches – two days, and so on. In time they would discover that there are always thirty days between one full moon and

Incas of Peru knotted cords called quipus for keeping count.

When man learned to herd cattle into natural pens and to sow and reap grain, he could stop wandering and live in a fixed home.

the next. So they might cut a bigger notch to mark a full moon. Twelve of these bigger notches would round off 360 days – roughly a year. We then have our first crude moon-calendar embracing the four seasons from spring to spring again.

After many thousands of years, some of these early hunters slowly began a new way of life. On returning to an old camping site, they would notice that grain left littered on their last visit was now sprouting in plenty. From this experience they learned to set some aside for planting. With the help of their constant companion, the dog, they also began to herd sheep, goats and cattle into ravines where it was easy to keep them penned in ready for slaughter only when there was need for meat. Instead of searching for wild herbs and berries, they sowed and reaped their own crops. They thus became shepherds and farmers.

As they settled down in villages they collected more and more goods which they could call their own. With hoes and digging-sticks, fields and fences, crops and herds, men needed to keep a record of their possessions. The earliest way of recording was the tally-system of the calendar-makers – one mark for one thing, two marks for two things, and so on. Counting this way lasted over a long period. In the New World, the Incas of Peru used to tie one knot in a cord to record each sheaf of grain gathered in at harvest, and in parts of the Old World there are still shepherds who cut chips in a stick when counting their flocks.

As men became farmers, they had to be able to forecast accurately the times of lambing and calving, of sowing and reaping. The hunter's rough-and-ready moon-calendar was no longer good enough. Nor was his way of recording numbers.

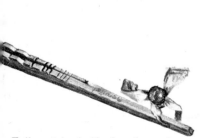
Tally stick: half of it formed a receipt in the Middle Ages.

Even today some European shepherds notch sticks when counting: one notch for ten sheep.

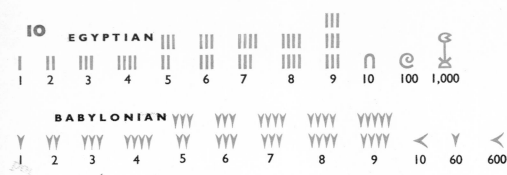

Early numbers show traces of notch-recording. Egyptians, Babylonians and Romans used strokes for the first few numbers and different signs for higher numbers.

of drought or storm and made thankofferings to the gods of harvest and abundance.

Though they thus mixed magic with their calendar-making, they did their job with surprising skill. Day by day they noted how the sun's rising position changed throughout the seasons; night by night they marked which star-clusters shone in the western sky where the sun had just set. In time they measured the length of the year to within an hour or two accurately. Without written records they could never have remembered all that their careful work taught them.

The earliest written numbers we know of were used in Egypt and Mesopotamia about five thousand years ago. Although these two lands are many miles apart, both their number systems seem to have started in the same way, by chipping notches on wood or stone to record the passing days. The priests of Egypt wrote on papyrus made from reeds, those of Mesopotamia on soft clay. So the shapes of their numbers are naturally different; but both used simple strokes for ones and different marks for tens and higher numbers.

If the farmer uses a moon-calendar of 360 days to forecast the seasons, he will make an error of five days the first year, ten days the next year and so on. Thus the wise men who were able to work out a sun-calendar, which is accurate, became people of special importance. Farmers willingly provided them with a living, so that they could devote their time to foretelling the seasons.

As time passed, the calendar-specialists became a ruling class. More often than not they were also priests, who offered sacrifices to appease the gods

The Mayas of Central America, cut off from the Old World, developed farming, building and time-reckoning in their own way.

MAYA

| 1 | 2 | 3 | 4 | 5 | 6 | 7 | 8 | 9 | 10 |
| 20 | 40 | 60 | 80 | 100 | 120 | 140 | 160 | 180 | 200 |

Just as our 0 makes a number ten times larger, Maya ⌒ made it twenty times larger.

Both built up the number they wanted simply by repeating the strokes and marks as often as necessary.

Three thousand years later the Romans still made strokes for the numbers one to four. They used new signs, in the form of letters, for fives, tens, fifties and so on. At about the same time, the people of China used a different sign for every number up to ten but still used strokes for the first three numbers.

The most remarkable of all early number systems was that used by the Mayas of Central America. Completely cut off from the civilisations of the Old World, these people could write any number with the help of only three signs – a dot, a stroke and a kind of oval. With dots and strokes only, they could build up any number from one to nineteen (≡). By adding one oval below any number, they made it twenty times larger, thus: •=1; ⌒=20. Adding a second oval would again multiply the number by twenty. In time-reckoning, however, they adjusted this system: adding a second oval multiplied the number by eighteen instead of twenty, so that ⌒ meant *not* 400 (1 × 20 × 20) but 360 (1 × 20 × 18). If we recall the moon-calendar of 360 days, we can understand why they used their number signs in this way.

In time the Mayas used a sun-calendar of 365 days. For their records of dates, carved on stone columns called steles, they used special numerals shaped like human faces.

Maya stele.

Taxes and Triangles

THE PRIESTS of early Egypt became the most powerful men in the land. It was they who fixed the many holy days – days when the feasts of the full moon or of the midsummer sun were to be held, days when animals sacred to certain star-clusters were to be sacrificed, days when offerings were to be made to the gods of the river. It was they, too, who ordered the building of great temples, which they also used as observatories, and of the mighty pyramids, which served as the tombs for their rulers, the Pharaohs.

To erect such stupendous buildings, the architects of Egypt had to know how to make some kind of ground-plan, how to level the edges of stone blocks, how to haul them up from the ground, and how to set them fairly and squarely in position. In learning all this, the pyramid architects were making practical discoveries in the art of measurement, or as we now say, geometry.

The first ground-plans, forerunners of our own blueprints, were probably drawn on clay, simple diagrams to show the shape of the finished building. Those who made them had indeed learned that two things – a drawing and a building – may be of quite different size but of exactly the same shape. So what is true about the shape of one is also true about the shape of the other.

When the ground-plan was complete, men with hoes levelled off a stretch of land ready for the masons to begin work. There were as yet neither wheeled vehicles nor good roads. Cargoes of heavy building materials, consisting mainly of stone blocks weighing several tons, came as near to the site as possible along the River Nile by boat.

Each block of stone had to be cut to shape. First the rough edges were knocked off with lumps of flint. Next the surfaces were levelled with metal chisels and bell-shaped wooden mallets. Last of all the whole block was smoothed by rubbing with a rough stone tool. Every corner had to be tested with a mason's square, or set-square, to make sure it was a true right-angle.

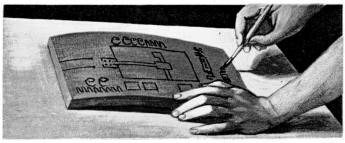
Scale plans were drawn on clay tablets before building began.

Then an enormous layer of blocks was laid to form the base of the pyramid. On this a second layer was built, slightly smaller and exactly in the middle of the first. Layer after layer was added in the same way so that all four sides of the finished pyramid would taper equally and meet neatly at the top. To check that it was upright, the edge of each had to be tested with a weight hanging from a string. Earth was piled up the step-like edges to make a sloping road over which the blocks were hauled on sledges with rollers beneath them.

Each block was shaped and corner tested with mason's square.

Heavy blocks were hauled up sand ramps on sledges, over rollers.

Plumb-line checked that the blocks were set exactly upright.

Perhaps the hardest problem was to make the base of the pyramid really square. The smallest error in fixing the angle at any corner would have thrown the whole building out of shape. Although the builders left no records, we may guess how they would do this.

They might mark out a long straight line, by stretching a cord between two pegs stuck in the ground. Then to each peg they would tie an equal length of string, more than half as long as the line they had drawn. By keeping these strings stretched tight and moving the ends around, they could draw parts of two perfect circles. These part-circles we call arcs will cross each other at two points. When the builder draws a straight line between these two points, he will find it bisects the original line, that is it crosses it at a right-angle, cutting it into two equal parts.

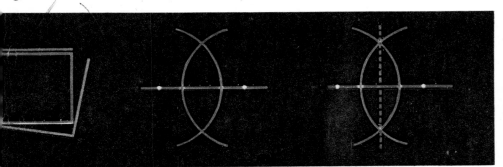
One false angle at base would ruin the shape. Right-angles can be made by drawing equal arcs from any two spots on a straight line, and joining the points where the arcs cross.

The builders of ancient Egypt checked that their walls were built at right-angles to the ground by means of a plumb-line.

The builder must be able to mark out right-angles on the flat ground to get his foundations square. To test whether his walls are dead upright, he also needs to make right-angles in the air. For this purpose, the Egyptian builders used the plumb-line, a device we still use today. If the plumb-line is suspended from the top of a wall so that the weight is free to swing, it traces an arc of a circle, and comes to rest at right-angles to the ground. If level with it, the wall is vertical.

When plumb-line stops swinging, it grazes the ground at right angles.

Of all ways of drawing a right-angle, the simplest is to use a set-square. The Egyptians did so. But first they had to *make* one, and to do this they had first to make a right-angled triangle.

Who first made this discovery we may never know. Possibly it was the professional rope-knotters whose job it was to tie equally-spaced knots in the long ropes used for measuring. Somehow they found that pegging out certain lengths of rope in the form of a triangle produces a right-angle opposite the longest side. Taking a length as the space between two knots, one combination that gives this result is 3 lengths, 4 lengths and 5 lengths. Another is 5, 12 and 13. By cutting pieces of wood of such lengths and by joining them end to end, they could make a set-square.

Crude ways of measuring, good enough for their forefathers, were not good enough for these builders of great temples and pyramids. The farmer who set out to build a stone or wooden hut with his own hands could say: My hut will

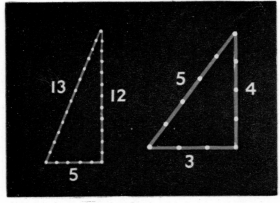 

A triangle with sides 3, 4 and 5 lengths has a right-angle opposite its longest side: so has one with sides 5, 12 and 13 lengths.

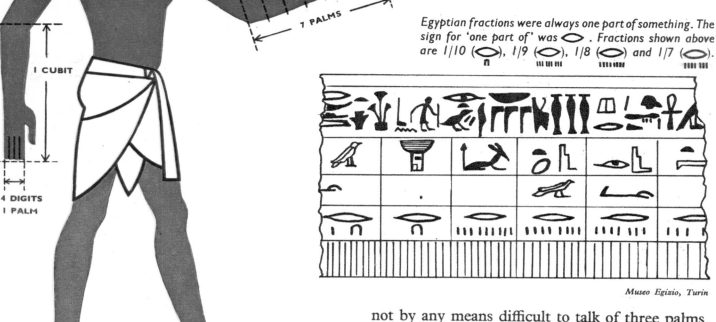

Early measures were based on the proportions of a man's body.

Part of an Egyptian measure, with marks for fractions.

Egyptian fractions were always one part of something. The sign for 'one part of' was ⬭. Fractions shown above are 1/10 (⬭), 1/9 (⬭), 1/8 (⬭) and 1/7 (⬭).

Museo Egizio, Turin

be six paces long and four paces wide, the roof will be a hand-span higher than the crown of my head. The temple architect could not give building instructions in paces and spans. Every workman under him might have a different pace and span.

For large-scale building there had thus to be measures that were always the same, no matter who did the measuring. In the beginning they were commonly based on the proportions of one man's body, possibly a king's. These standard measures were fixed by rulers of wood or metal.

In Egypt the main standard of length was the cubit, often mentioned in the Bible. It was the length of a man's forearm from the elbow to the tip of the outstretched middle finger. There were also smaller measures: the palm, one-seventh of a cubit, and the digit, one-quarter of a palm.

Such smaller units were very important to the Egyptians because they found fractions hard to handle. We now think easily of fractions such as three-fifths, or nine-tenths, but a fraction to an Egyptian was always one part of something. It was too hard to think of three-sevenths of a cubit but not by any means difficult to talk of three palms.

These early measures seem strange at first, but equally strange ones are still in use. A Briton or an American still measures his own height in feet. He still says 'seven inches' to avoid using the fraction 'seven-twelfths of a foot'.

At harvest time each year the priests of Egypt levied payment for their services by collecting taxes from the farmers who paid in goods. To fix the payment due, there had to be standard jars for measuring out grain, wine or oil, and standard weights for weighing other produce.

Scales were used for weighing, jars for measuring.

British Museum

Before gathering taxes the priests measured the area of each field. Odd-shaped fields were first marked off into triangles.

It seems that the amount of tax depended on the size of the farm; the bigger the farm, the bigger the tax. To levy taxes, the priests therefore needed some way of measuring area.

Perhaps their first clue to area-measurement came when paving the floor of a temple with square tiles. A strip of floor six tiles long and six wide needs thirty-six tiles (6 × 6) to cover it. Another strip, ten tiles long and four wide, needs forty (10 × 4). To find the area of a square or oblong, you merely multiply its length by its width.

But not all fields were square or oblong. The tax-gatherer would come upon fields that seemed all sides and corners. There was no way of dividing them into squares, but he *could* easily divide them into triangles. If he knew how to find the

Base 3, height 3. Area 9.

A square folds into two triangles. Each is half the area of the square.

Base 5, height 3. Area 15.

An oblong cuts into two triangles. Each is half the area of oblong.

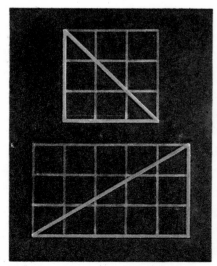
The area of any right-angled triangle is half the area of the square or oblong with the same base and height.

area of a triangle he could thus measure *any* field, providing its sides were all straight.

Happily it is only a short step to learn how to find the area of a triangle once you know how to find the area of a square or an oblong. A square piece of linen will fold into two equal triangles, each half the size of the square. An oblong piece will cut into two equal triangles, each half the size of the oblong. Possibly such simple clues gave the priests a guide to the rule they needed. They saw that we find the area of a triangle by multiplying its base (or length) by its height (or width) and dividing the answer by two.

The job of measuring the fields kept the priests busier than we might think. The farmland of Egypt lies in a narrow strip near the great River Nile. On either side of this fertile strip, all is desert. At midsummer each year the river overflows its banks, watering the nearby land and leaving behind a thin layer of rich, muddy soil when the floods go down again. This yearly flooding helped the early farmers of Egypt to

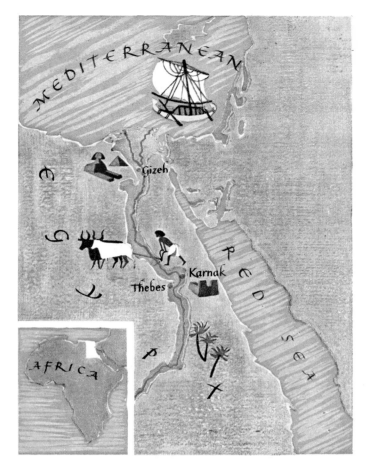

Food grows only where the Nile floods. Beyond, all is desert.

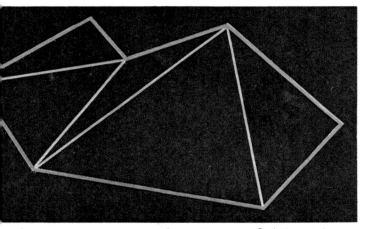

By measuring every triangular strip we can find the total area.

grow fine crops, but it also washed away the boundary-marks between their fields. So the priests had to measure each plot of land again and again, year in and year out. They were not merely calendar-makers and architects. They were also the world's first professional surveyors.

Ever since the days of the Egyptians, the main working method used by surveyors of all ages has been what they call triangulation. When we think about how much the early priest-surveyors must have learned about the shapes and areas of triangles, we begin to see how much the mathematicians of later times gained from their practical knowledge. To it, we owe the very word geometry, made up from two Greek words, one meaning earth or land and the other measurement.

Of course, the surveyor meets problems that he cannot solve by the simple rule for finding the area of the triangle. He cannot mark out a circle in strips which are exactly triangular.

The early Egyptians almost certainly drew circles by pulling a tightly stretched cord around a fixed peg. They knew they must use a long cord to draw a large circle and a short cord to draw a small one. They knew, in fact, that the area of a circle depends on the distance from its middle-point to its edge, or on what we now call its radius.

About 3,500 years ago, when the great pyramids were already very old, an Egyptian scribe named Ahmes, the Moonborn, put a rule about this in writing. The area of a circle is very nearly three and one-seventh times as great as the area of a square drawn on its radius. That is, if the radius is 3 inches, the area of the circle is roughly $3\frac{1}{7} \times 9$ square inches. How the priests made this discovery we may never know. In the British Museum in London hangs the carefully framed papyrus manuscript written by Ahmes himself. Unfortunately it gives us no explanation.

Symbol for radiance

Finding North and South

North of the Tropic of Cancer, the sun's noon shadow points always due north.

Finding East and West

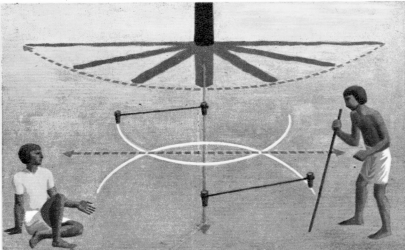

A line drawn at right-angles to the noon shadow points due east and west.

Ra, God of the Sun, holds the sceptres of east and west.

The temple at Karnak was built in such a way that at dawn on midsummer's day a man looking along the line of columns would see the rising sun straight ahead.

At dawn this Egyptian sundial was set with its crossbar facing towards the east. The longest shadow marked the sixth hour before noon. At mid-day, when the shadow was shortest, the dial was turned round. Slowly the shadow lengthened and marked the afternoon hours.

Scattered in museums around the world there are other early manuscripts which give us a glimpse into the mathematics of Egypt, but most of our knowledge comes from examining the ancient buildings which still stand near the Nile.

We can tell how accurately the priestly architects could fix direction from the fact that the four faces of certain pyramids look precisely toward north, south, east and west. The architects probably found north and south from the noon shadow of some tall column. By drawing a line at right-angles to such a shadow, they could place east and west as well.

There is also another way of finding east which the Egyptians must have understood. Day by day, the sun's rising position gradually changes. In winter it rises to the south of east, in summer to the north of east. If you can halve the angle between its midwinter and midsummer rising positions you will know what is due east.

By counting the days between two occasions when the sun reached its most northerly rising position, the Egyptians could measure the length of the year. At Karnak they built a temple with a line of columns pointing to where the sun rose on midsummer's day. Only once in 365 days did the rising sun shine straight along that line.

In finding direction and measuring time, the Egyptian had only the same clues as the hunters and food-gatherers of a bygone age: the rising and setting positions of sun, moon and stars, the shadow of the sun by day and the rotation of star-clusters around the Pole Star at night. Years of careful recording, however, enabled the Egyptian to make far better use of these clues. The early hunter looking at the long shadow cast by a tree could say at best: It is still early morning. The Egyptian, with a sun-clock which measured the length of a shadow falling on a marked strip of wood, could look at the shadow and say: The second hour of morning is at hand.

Here we have real science; but many of the priestly drawings of ancient Egypt show the gods busy controlling the points of the compass or the hours of day and night. Along with real science they trailed a heavy load of superstition.

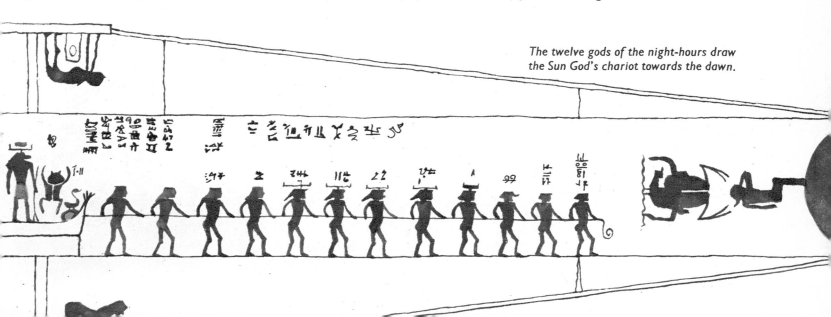

The twelve gods of the night-hours draw the Sun God's chariot towards the dawn.

Square and Circle

A THOUSAND miles east of the delta of the River Nile are two other great rivers, the Tigris and Euphrates. Between and beside their banks, in the land called Mesopotamia, there grew up another civilisation, at least as ancient as that of Egypt.

Historians refer to this civilisation at different stages of its development as Sumerian, Chaldean, Assyrian and Babylonian. It was in some ways very similar to the Egyptian. In both, the priestly sky-watchers and calendar-keepers were the ruling class and both made astonishing progress in astronomy. By about 2000 B.C. the priests in these lands had built up temple-libraries where they recorded their knowledge in a secret script which ordinary men could not read.

There the resemblance between the two civilisations ends.

Mesopotamia, unlike early Egypt, had a considerable foreign trade. It had no wood of its own suitable for building, no silk of its own to clothe kings and princes, no spices for the dishes of the wealthy, few precious metals from which to make vessels for the temples. To meet all these needs, merchants with caravans of asses or camels travelled through mountain pass and over desert, going westward to Lebanon for cedar-wood, northward into Asia Minor for gold, silver, lead or copper, and eastward possibly as far as India and China for silks, dyes, spices and jewels.

A merchant, selling the produce of his fields, may be content to measure his wares roughly and to sell them by the donkey-load; but a merchant dealing in more costly goods needs to be far more precise.

Thus, in Mesopotamia, scales and standard weights came into common use, and the merchant weighed his heavy goods in talents (roughly fifty-five pounds) and his precious wares in shekels (rather less than one-third of an ounce). But the merchant also needed to find something which anyone and everyone would accept as payment for goods. There was one thing which almost everybody would accept: barley. For many

Later, the merchants learned to use scales and standard weights.

Early Babylonians weighed and sold their goods by the donkey-load.

Still later, standard weights of silver were used as money.

Assyrian lion weights.

years, barley was the workman's wage. What was left over after he had made his bread and brewed his beer he could exchange for other things. So the early Mesopotamian merchants, when they set off to trade with other lands, loaded their asses and camels with barley to pay for the goods they intended to buy.

As time went on they discovered that silver, much lighter and easier to handle than barley, would also be acceptable almost everywhere. At first, they would carry small quantities of it and weigh it out as necessary. Later, they side-stepped the constant trouble of weighing by casting small bars of silver each stamped to show its weight. Although not much like our modern coins in appearance, they were the world's first money.

Here, for the first time, was a kind of wealth which a man could save without fear that it would go bad. He could also lend it out and charge interest on it, as the usurers of the Bible story did. To do this, as when buying or selling, he would need to keep accounts.

In this task, the merchants of Mesopotamia were handicapped by a clumsy script and bulky writing material. They wrote with pointed sticks on tablets of soft clay; and they had to bake the tablets hard in the sun to hold the impression. The process must have slowed down writing considerably, but it made the finished tablets difficult to destroy. In recent years archaeologists have found thousands of them with wedge-shaped signs, called cuneiform, still clearly written on them.

It needed masterly detective-work to decipher the writing, partly because the signs, at first sight, all look very much alike, partly because different scribes used signs in different ways.

What stood for 10 in one place might stand for 60 in another; what stood for 100 (10 × 10) on one tablet might mean 3,600 (60 × 60) on the next.

Although Mesopotamia had elaborate systems of weights and measures, and although it was the first home of money, its methods of keeping written accounts remained at a very crude level. Fortunately for them, however, the merchants had a way of calculating without written numbers.

Babylonian duck weights.

Between two great rivers lies cradle of many civilisations.

The early scribes did their calculations on an abacus—pebbles in grooves in the sand. They recorded the results on clay tablets.

Like the Egyptians, they set out pebbles in grooves in the sand, each pebble in the first groove standing for one, each in the second groove for ten, each in the third for a hundred, and so on. The diagrams below show how the merchants used this device, called the abacus, for adding up their accounts, long before there were any rules for written arithmetic.

Notice how the number-value of a pebble grows as you move it from groove to groove: 1 in the first groove, 10 × 1 in the second, 10 × 10 × 1 in the third, 10 × 10 × 10 × 1 in the fourth. When

and so on, repeated to recall the number of pebbles in the corresponding groove. Although the people of Mesopotamia also used a base of ten, they sometimes used a base of sixty. A trace of this has come down to us. In measuring time, we still divide the hour into sixty minutes and the minute into sixty seconds. Navigators, in measuring distance, still divide each degree of longitude or latitude into sixty minutes, and each minute into sixty seconds. The Mayas of Central America, who may have used both toes and fingers for counting, used a base of twenty except in time-reckoning.

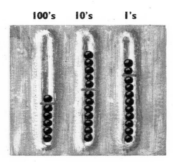

To add 579 to 152, first place pebbles to show 579: 5 hundreds, 7 tens, 9 ones.

Add pebbles to show 152: 1 hundred, 5 tens, 2 ones. Ones column has 11 pebbles.

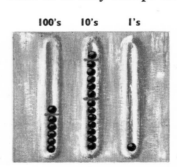

Carry 1 from ones column to tens, throw 9 out, leave 1. Tens column now has 13.

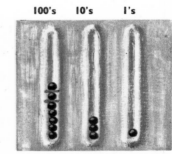

Carry 1 from tens to hundreds, throw 9 out, leave 3. Pebbles now give answer, 731.

the value of a pebble in each groove or column is ten times greater than in the one before it, we now say that ten is the base.

Most ancient number systems used a base of ten, probably because most people first learned to count on the fingers of their two hands. But there is nothing magic about the number ten. It is just as easy to work with a quite different base.

The Egyptians used ten as a base, and consistently used separate signs for 1, 10, 100, 1000

Mesopotamia was not the only country handicapped by clumsy signs for numbers. The same thing was true of most civilised lands until a few centuries ago, and so the habit of using the abacus spread in course of time over most of the world.

The abacus used in ancient Rome was a metal plate with two sets of parallel grooves, one below the other. The lower set held four pebbles in each groove and the upper set only one. The pebble in an upper groove was worth five times as much as a pebble in the corresponding lower groove. Thus the operator could show any number up to nine in each complete column. At the right of the metal plate there was a separate set of grooves used for working with fractions. The word the Romans used for a pebble was *calculus*, from which we get our own word calculate.

The abacus was not the only short-cut to calculation known to the merchants and traders of Mesopotamia. Among the thousands of clay tablets which archaeologists have unearthed from a temple library near the banks of the Euphrates, some are tables of multiplication and addition, and others are tables of the squares of numbers. The

Japanese businessmen still use the abacus with great skill.

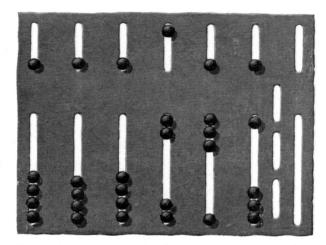
Roman bronze abacus. Beads in the top row count five.

square of a number simply means a number multiplied by itself, such as $2 \times 2 = 4$ which we now write as 2^2, or $5 \times 5 = 25$, which we now write as 5^2.

It seems likely that the priests of those days had discovered a way of using square tables which enabled them to multiply any two numbers together without using the abacus. Here, for example, is how they would multiply 102 by 96.

STEP 1 Add 102 to 96 and divide the result by 2 to find the average..............99
STEP 2 Take 96 from 102; divide the result by 2, to find half the difference between the two numbers.....................3
STEP 3 Look up in the table the square of 99 and you at once see it is....9801
STEP 4 Look up in the table the square of 3 and you at once see it is..........9
STEP 5 Take 9 from 9801 and you find the correct answer.....................9792

If we understand this method, we can multiply any two numbers together in the same way. When we multiply one number by another, the result is always equal to the square of their average minus the square of half the difference between them. Yet square-table multiplication was never as widespread as the use of the abacus. Long after the time of Columbus, some merchants and shop-keepers of western Europe still used counting-boards worked on much the same principle as the abacus. The modern Chinese, Japanese and Russian businessman frequently uses the abacus today and works with great speed.

Square table shows the results of multiplying numbers by themselves.

Part of the above, Babylonian style.

Wheel helped the potter to work faster and better.

Mesopotamia found other uses for the wheel, too. By placing his clay on a turning wheel, the potter could mould his vessels more accurately. With the help of pulley-wheels, builders and engineers could raise heavy weights more easily.

It is tempting to suppose that the Mesopotamians, with their knowledge of wheels, learned a good deal about the geometry of the circle; but in fact they made no more progress than the Egyptians—probably not as much. The Egyptians estimated that the boundary, or circumference, of a circle is 3.14 times as long as its diameter. Since this ratio of circumference to diameter (or of area to square of radius), now called by the Greek letter π (prounced pi) is approximately 3.1416, this was quite a close estimate. The Mesopotamians were

The pulley-wheel made it possible to raise heavy weights.

About six thousand years ago some unknown citizen of ancient Mesopotamia made one of the greatest inventions of all time, the wheel. At first it was no more than a solid disc of wood with a hole in the middle to allow it to revolve round a fixed axle. By the time the Babylonians and Assyrians built their trading carts and war chariots it had become much more like the wheel of the farm cart still seen today, with rim, spokes and hub.

commonly content to use the more convenient but less precise value, 3.0.

How these early peoples arrived at any value for π, however crude, is not certain; but some of their inscriptions give us a clue. By drawing the smallest square that can enclose a circle and the largest that can fit inside it, they could see that the boundary of a circle lies between the boundaries of the two squares, and it happens that the average boundary

of the two squares is almost $3\frac{1}{7}$ times the diameter of the circle. By drawing one hexagon outside, and another inside the circle, they could have obtained an even better estimate.

If the priests of Mesopotamia knew less about the circle than did those of Egypt, their knowledge of practical geometry was not inferior. In astronomy, also, they were as far advanced as the Egyptians. A work on astrology, prepared for Sargon, King of Babylon, almost five thousand years ago, includes a long list of the times of eclipses.

It is easy to understand why the sky-watchers of those days were so interested in eclipses. Astrology was a strange mixture of science and magic. The priests claimed that they could foretell from their observation of the heavens all kinds of things—the outcome of battles, the fortunes of kings, the wrath of the gods. If the priest could forecast eclipses with accuracy, people were more ready to listen to other prophecies he might make and to give him more power.

We now know that an eclipse of the moon occurs when the earth is in a straight line between sun and moon. The earth then casts a shadow across the moon's face. The priests of Mesopotamia forecast eclipses of the moon with sufficient reliability to make us think that they too probably knew this.

The astronomers of Babylon were keen observers of eclipses.

Babylonian map shows the earth as a disc.

British Museum

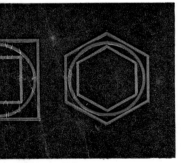

Half sum of outer and inner figures gives size of circle roughly.

This clay tablet shows early interest in square and circle.

Looking up at the round edge of the shadow on the partly-eclipsed moon, they would then realise that the earth itself must be round. Babylonian scribes did indeed draw fanciful maps of an earth whose shape was like a penny. On one such map which archaeologists have found in recent times, Babylon occupies a large area in the middle.

They sailed south to the coast of western Africa for spices.

Stars and Steering

AROUND 1500 B.C. a large sea trade sprang up along the coast of Syria, in the land the Bible calls Phoenicia. From the great ports of Tyre and Sidon, Phoenician seamen sailed the length and breadth of the Mediterranean, fetching and carrying goods for most of the known world. The Old Testament prophet, Ezekiel, gives a picture of Tyre at the height of its power: Tyrus, thou art situate at the entry of the sea, a merchant of the people for many isles. Thy builders have made all thy ship boards of fir trees of Senir: they have taken cedars from Lebanon to make masts for thee. Fine linen with broidered work from Egypt was thy sail. Tarshish was thy merchant by reason of all kind of riches; with silver, iron, tin and lead they traded in thy fairs.

The Phoenicians ventured northwards to islands off Britain for tin.

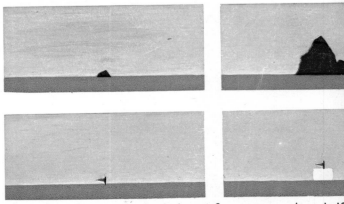

Far off, tip of ship or island shows; from nearer, about half

Nearer their homeland, the merchant-mariners went to Greece for cargoes of wine and olives; to Egypt for grain and linen.

With so many people flocking to buy their goods, ambitious sea-traders were tempted to send fleets on even longer voyages. By 1000 B.C. their ships had probably ventured into the Atlantic for tin from the Isles of Scilly, near Britain, and southward along the coast of Africa for spices.

In the course of their travels, the Phoenician mariner-merchants met the less advanced folk of Europe and of the Atlantic coast of Africa as well as the more civilised peoples of Egypt and of Mesopotamia. Their ships that plied the seas carried from one place to another valuable knowledge as well as valuable cargoes; and the long voyages also taught them much that the Egyptians and Mesopotamians had no opportunity of learning about the earth and the heavens.

People who live on or near the sea soon grow familiar with notions which would never occur to those who live inland. The citizen of Babylon or Nineveh would probably have laughed if an astrologer had tried to convince him that the earth is spherical; but the men of Tyre and Sidon could find it out for themselves.

A merchant watching for a ship to come into port, would first see the tip of a mast above the horizon, then the top half of a sail, and last of all the whole ship. A ship's lookout, if watching for some nearby island, would first see the top of its highest hill, then the lower slopes and finally the shore as well. It would be hard for either to explain what each saw without guessing that the curved surface of the earth hides the lower parts of distant objects from view.

Like most seafarers before the days of Columbus, the early Phoenicians seldom sailed farther than needs be from sight of land or familiar land-birds. In the long, narrow Mediterranean they could undertake east-west voyages of up to two thousand miles without long losing sight of a familiar coast or island. While the pilot could recognise such landmarks, he needed little knowledge of navigation as we now use the word.

Close up, whole is seen.

Thus the mariners learned that part of a distant view is hidden by the curved surface of the earth.

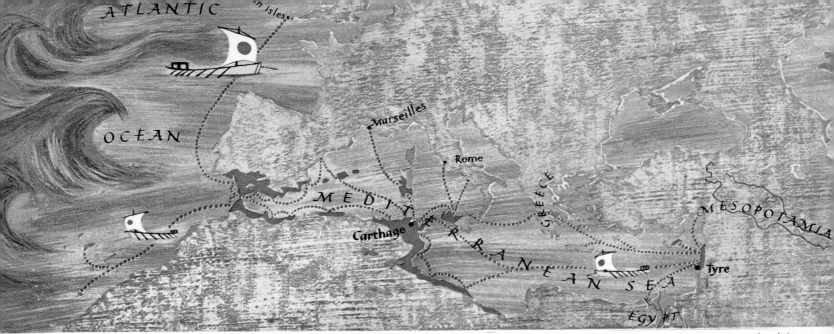

Phoenicians founded colonies all along the Mediterranean. From western ports they made north and south voyages in the Atlantic.

In time the Phoenicians founded colonies near the western end of the Mediterranean, including the great city-port of Carthage which became so powerful as to challenge the might of Rome to war three times in a single century. From these western colonies, mariners set out on voyages into the Atlantic. Sailing by strange shores with no familiar landmarks to guide them, they needed some way of reckoning how far north or south they had travelled.

Like the early hunters, they had to rely on the sun and on the stars; but they began with the advantage of already knowing something about astronomy. They soon learned more. Once pilots pushed out into the Atlantic and voyaged northward for tin, southward for spices, an entirely new picture of the changing heavens met their eyes.

Along the European coast the noon sun, on any particular day, is lower in the sky at northern ports than at southern ones, and casts a longer shadow. The length of the noon shadow, different at different places on the same day of the year, gave the mariner his first daytime method of marking the latitude of a port on his crude map or chart. By night, a star close to where our North Pole Star now lies would equip him with another method. As a pilot sailed northward towards the Tin Isles he would see this star rise a little higher in the sky each night. As he sailed southward along the African coast he would see it dip night by night a little nearer to the horizon.

From earliest times men have noticed how sunbeam, moonbeam or starlight forms parallel straight lines. Slowly, the Phoenicians would

Noon shadows, long in the north, grow gradually shorter as one nears the Equator.

From very early times men have noticed how the rays of the sun are like parallel lines.

connect their newly gained experience of the heavens in different latitudes with this age-old knowledge and realise that only one simple explanation fits the facts. That our earth is spherical can explain why such beams strike various parts of it at different angles at the same moment.

In northern seas the Pole Star is seen high overhead. Nearer to the Equator it seems to dip lower and lower towards the horizon. This changing angle gives the pilot a clue to his ship's north-south position.

Other discoveries they made in their travels would strengthen this belief. In their own land the sun's noon shadow always points to the north, but on the Tropic of Cancer (about $23\frac{1}{2}$ degrees north) they would see the noonday sun at midsummer directly overhead, casting no shadow at all. Still farther south on the same day, the noon shadows point to the south.

On rare occasions, a captain would sail his ship far south along the coast of Africa. As he neared the Equator, he would see a star near the North Pole of the heavens very low in the sky. Farther south still he would lose sight of it altogether. Instead he would see bright stars and clusters which are never visible in northern lands.

The early pilots, with their new experience of the heavens, blazed a trail for a new science of navigation leading to further advances in the study of earth-measurement or geometry. During the early centuries of Mesopotamian civilisation, long before the time of the Phoenicians, man had learnt to divide the circle into 360 degrees. Now man was learning to divide the great circles which pass through the North and South Poles of the earth itself in the same way.

The mariner-merchants of Phoenicia and Carthage led the world in navigation for many years. As time went by Greek-speaking seamen of Sicily, Crete, Cyprus and many of the isles off the coast of Asia Minor, as well as what we now call Greece, challenged their leadership. By about 400 B.C. Greek geographers were beginning to draw charts on which the Mediterranean coastlines are recognisable.

But the Phoenicians left the Greeks a legacy far more valuable than such crude map-making. The men of Tyre and Sidon, who spoke a language rather like Hebrew, were among the earliest people to use a new sort of writing. Instead of using a vast number of picture-symbols for words or ideas, the Phoenicians used an alphabet made up of a few simple signs which stand for sounds. About 600 B.C. the Greeks adapted such an alphabet to their own very different language. Thereafter it was an easy matter to master the art of reading and the written word was no longer a secret.

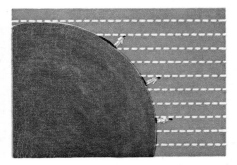

The angle of the sun's rays to the curved earth affects the length of the shadow.

By observing length of noon shadow, navigator can locate ship's position on chart.

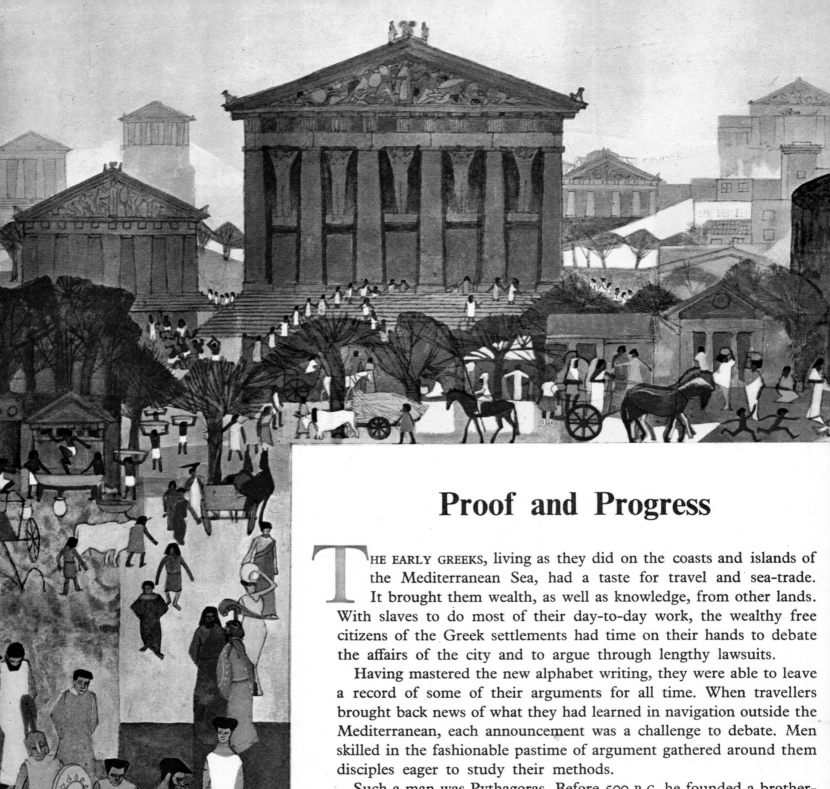

Proof and Progress

THE EARLY GREEKS, living as they did on the coasts and islands of the Mediterranean Sea, had a taste for travel and sea-trade. It brought them wealth, as well as knowledge, from other lands. With slaves to do most of their day-to-day work, the wealthy free citizens of the Greek settlements had time on their hands to debate the affairs of the city and to argue through lengthy lawsuits.

Having mastered the new alphabet writing, they were able to leave a record of some of their arguments for all time. When travellers brought back news of what they had learned in navigation outside the Mediterranean, each announcement was a challenge to debate. Men skilled in the fashionable pastime of argument gathered around them disciples eager to study their methods.

Such a man was Pythagoras. Before 500 B.C. he founded a brotherhood of young men to whom he imparted his mathematical knowledge only after they had sworn an oath never to pass it on to an outsider. Despite this secrecy and despite the fact that he mixed magic and religion with such instruction, Pythagoras was thus a pioneer of the teaching of mathematics. A century later there were, in Athens, schools where philosophers such as Plato taught young men law, politics, public speaking and mathematics. In these new schools there were no oaths of secrecy. Teachers and pupils were free to put whatever they wished in writing for all to read.

Public study and debate led to a new way of thinking about mathematics. While the peoples of the ancient world knew many useful rules for finding areas and measuring angles, they had never attempted

Plato

Over the door of his school was written:

Let no one ignorant of geometry enter.

Sea-trading brought both wealth and knowledge to the people of the Greek islands.

to link those rules in a train of reasoning to prove that they are reliable. The argumentative Greeks insisted on putting every rule to the test of debate and answering every objection brought against it.

Long before Pythagoras, it was known that a triangle with sides 3, 4, and 5 lengths is a right-angled triangle and that the same is also true of a triangle with sides 5, 12 and 13 lengths. But Pythagoras noticed something common to both sets of numbers. If we look at the squares of the numbers in each set we see that the square of the largest number is in each case equal to the sum of the squares of the two smaller numbers:

$$3^2 = 9 \qquad 4^2 = 16 \qquad 5^2 = 25$$
$$5^2 = 25 \qquad 12^2 = 144 \qquad 13^2 = 169$$

Both these recipes for making a right-angled triangle mean that a square drawn on the longest side is equal in area to the sum of squares drawn on the two shorter sides.

Pythagoras was able to recognise the rule which embraces all such recipes. What is more interesting about him is that he asked himself two novel questions. First, is the rule *always* true? Second, is a triangle necessarily right-angled if its sides do obey this rule?

Pythagoras was not content with collecting more and more examples to show that the answer to both questions is yes for each example. He set himself the more ambitious and more useful task of proving that the answer must *always* be yes.

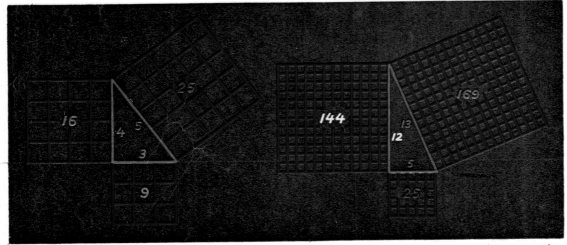

Square on longest side of a right-angled triangle equals sum of the squares on other two sides.

The free citizens of Greece who spent much time debating the affairs of their city-states, became expert in the art of argument.

To satisfy his followers that a rule was sound, the Greek teacher of mathematics, who was also a teacher of law, would have to argue as carefully and consistently as if he were fighting a case in the courts. Just as a judge and jury want to know exactly what a lawyer means when he talks about violence or negligence, the disciples would want to know precisely what the teacher meant when he spoke of figures, lines or angles. He had to give thoroughly satisfying definitions.

Finding a form of words to satisfy everyone was not easy. It raised a host of questions. How shall we define a straight line so as to be quite sure what we are talking about? How shall we define a circle clearly enough to stop the doubter from saying that he understood us to mean any round figure, such as a sphere or an oval?

As the argument continued, it became clear that the best way of defining a particular figure, such as the triangle or the regular hexagon, is, usually, to state how you make it with the means at your disposal. This raises yet another question: what means will everyone agree to put at your disposal? The only means the Greeks agreed upon were the ruler for making a straight line and the compass for drawing a circle.

The cautious Greeks would take very little for granted. They accepted on trust only definitions and ideas that seemed quite evident. Everything else had to be proved up to the hilt; but the teacher

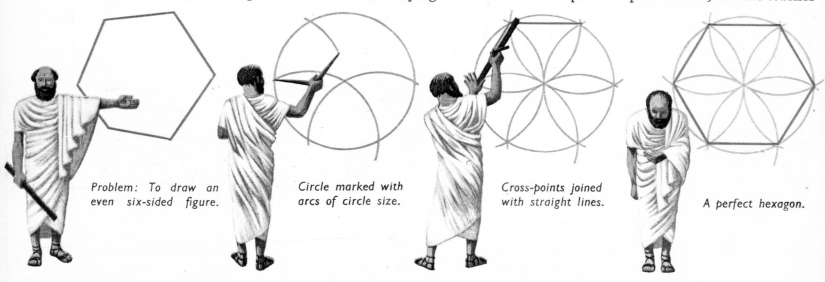

Problem: To draw an even six-sided figure.

Circle marked with arcs of circle size.

Cross-points joined with straight lines.

A perfect hexagon.

Euclid's general proof of Pythagoras's Theorem.

Euclid's work set the pattern for all that we mean by proving a case in mathematics.

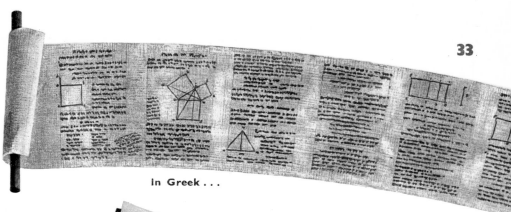

In Greek...

In Arabic...

In English...

who had thus proved that a new rule was sound could use the same rule when proving another. There was no need to start right at the beginning again. By arranging his proofs in order, with one rule leading to the next, he could save a lot of tiresome repetition. This was the way in which the great philosopher, Plato, taught the youth of Athens in the fourth century B.C.

Before Greek times there had been no logical system of rules—no science of geometry. There had been only disconnected recipes—for making angles of various sizes and triangles of different shape, for finding the area or circumference of a circle, and so forth. By the time of Plato it was possible to shape all these into an orderly and reasonable scheme of rules.

This vast scheme has come down to us through the writings of one man, Euclid of Alexandria. About 300 B.C. Euclid wrote a series of textbooks which have proved to be the best-sellers of all time. A thousand years later, when so many Greek writings had been lost or destroyed, Euclid's *Elements* were translated into Arabic and studied in the Moslem universities. Until fifty years ago translations into modern languages were used as textbooks in European and American schools. Even today school geometry is still mainly a streamlined version of the geometry of Euclid.

We now know that Euclid took for granted some things which need not necessarily be true when we use geometry in the service of astronomy. Indeed, Euclid's geometry is now only one of several possible systems. Yet his work still remains the model of everything we mean when we talk about proving a case in mathematics.

A Greek book in Euclid's time was a bundle of papyrus scrolls.

Treasury of Cnidus shows use of triangle and parallel lines.

The Greeks studied geometry less for any practical benefits it might bring them than for fun. Yet they found that the knowledge they gained proved useful in the world's work. It was useful for building, for navigation and for astronomy, for the layout of cities and for the design of musical instruments. As the science of land measurement it especially helped the surveyor.

Among the rules which the Greeks had mastered by the time of Pythagoras are these two: (1) the three angles of any triangle add up to two right angles (180°); (2) if two angles of a triangle are equal, the sides opposite to them will also be equal. From the first rule we can see that if one angle is a right-angle (90°) and another is half a right-angle (45°), the third must also be half a right-angle (45°). From the second rule we know that the sides opposite the two angles of 45° are equal.

When sunbeams strike the earth at an angle of 45°, a pillar, its shadow and the sunbeams form just such a triangle. This gives the surveyor a method of measuring the height of the pillar without the trouble of climbing it. Pillar and shadow are both opposite an angle of 45°, so both are of equal length. Instead of measuring the height of the pillar, the surveyor can measure the shadow.

When we know the rule about a triangle of this kind, we can apply it in many other ways. For example, we can tell how far out at sea a ship is if its course lies parallel to the shore. All we have to do is to find one point where a person can sight it at exactly a right-angle, and a second point where another can sight it, at the same time, at half a right-angle. The distance between these two points is the same as the distance of the ship from the shore.

Thales, the Greek master-pilot who was at one time the teacher of the young Pythagoras, boasted that he disclosed this trick of measuring to the priests of Egypt. More likely it was the other way round; but we can be sure that the priests could not give as good reasons for believing the rule to be true as could the logically-minded Greeks.

Compass and ruler were the only tools of Greek geometry.

Tower of Greek design, built in Roman times.

Measuring height: Sun's rays are parallel, so both these triangles have the same shape. The height of the stick bears the same relationship to the length of its shadow as the height of the pyramid bears to the length of its shadow measured from the middle of the base.

Thales probably knew a second rule for finding the height of a pyramid from its shadow; and the Egyptians probably knew the same recipe without being able to prove why the rule, which is the master-rule of surveying, is always true. In any two triangles whose corresponding angles are all the same, the ratio of the lengths of any corresponding pair of sides is also the same. Thus the height of a pole will have the same ratio to the length of its noon shadow as the height of the pyramid has to the length of *its* noon shadow added to half the width of its base.

The surveyor who knew only the rule applying to a triangle with two angles of 45° could not measure the height of a pyramid from its shadow except on certain fixed days in the year when the noon sun stood at just the correct angle in the heavens. Once he had learned the rule about corresponding triangles he could do so at any season of the year.

It was not until some three hundred years after the time of Thales that the Greeks accomplished their most interesting feat of surveying. The scene, once more, was set in the Nile Valley.

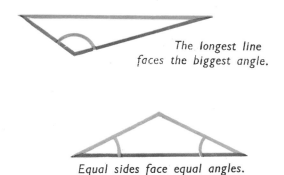

The longest line faces the biggest angle.

Equal sides face equal angles.

When one angle is a right-angle and two sides are equal, the equal sides each face half a right-angle.

Measuring distance: Ship is sighted at right-angle to shore and at half right-angle. Distance between sighting points is distance of ship.

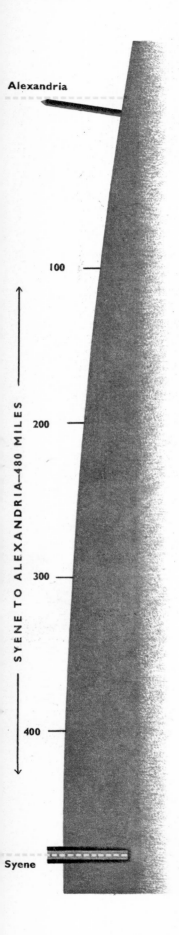

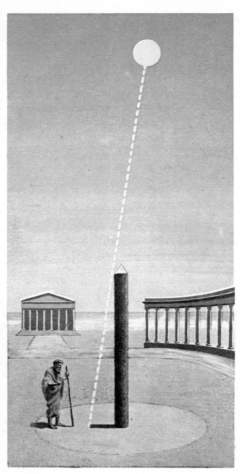

The noon sunbeam strikes Alexandria and Syene at different angles. This clue enabled Eratosthenes to measure the Earth's size.

After Alexander the Great, king of what is now Greece, had conquered Egypt (332 B.C.), the city of Alexandria, built by his orders and named after him, became the chief seat of learning in the Mediterranean.

One of the many brilliant mathematicians who taught in its schools was Eratosthenes. About 240 B.C., as librarian of Alexandria's already unsurpassed library of scrolls, he learned that Syene, near what is now called Aswan, stands almost exactly on the Tropic of Cancer. At noon the reflection of the midsummer sun was there visible in the water of a deep well. This showed that the sun was directly overhead and that its beams therefore pointed in a straight line towards the middle of the earth. On the same day, measurement of the noon shadow cast by a pillar at Alexandria shows that the sunbeam strikes the earth at an angle of 7 1/5° off the vertical. We know

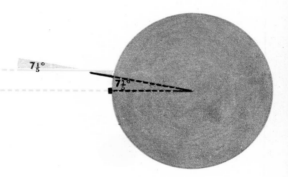

that sunbeams travel in parallel straight lines, so we may account for the difference only by the curve of the earth.

If we draw two parallel straight lines, one to show the sunbeam at Alexandria and the other to show the sunbeam at Syene, we see that the line on which the vertical pillar stands cuts through both of them. It cuts the first at the surface of the earth, and the second at the middle of the earth. The Greeks knew that when a straight line cuts across two parallel straight lines, it makes equal angles with both of them. Eratosthenes thus knew that the angle between Alexandria, the middle of the earth, and Syene must be 7 1/5°, which is exactly one-fiftieth of the 360° circle.

Syene lies nearly due south of Alexandria, and the road between them therefore lies almost exactly on a great circle passing through the North and the South Poles. Since it is almost exactly 480 miles long, the great circle is 50 times 480 miles in length. That is, the circumference of the earth is about 24,000 miles. Eratosthenes gave this remarkably accurate estimate of the size of our earth more than seventeen hundred years before Magellan's ships first sailed round it.

The Greeks of this period used mathematics in many ways we are apt to regard as modern. Archimedes, the greatest mathematician of the age, was also the inventor of many mechanical devices. He made a screw which revolved inside a tightly fitting cylinder, raising water as it turned. This was used for irrigation and for draining ships. While testing whether a crown was of pure gold or of mixed gold and silver, he discovered the principle of buoyancy. That is, a body plunged into fluid loses as much of its weight as will counterbalance the weight of the fluid it displaces. We use this principle today in making hydrometers to measure the density of liquids.

Among his many contributions to mathematics he was able to give a much more accurate value for π. We have seen that we can get a rough-and-ready value by taking the average of the boundaries or areas of the squares which just enclose and just fit inside the circle. We can cut down the possibility of error by narrowing the limits of the inside and outside figures. Our diagram shows how the limits are narrowed by using twelve-sided figures instead of squares. Archimedes went further. He used regular figures with 48 sides and was thus able to find a value which is still regarded as accurate enough for most practical purposes of design and engineering.

He also discovered how to find the volumes of various solid figures. Two of them, the sphere and the cylinder, were shown on his tombstone.

Archimedes uses water-test to check that crown is pure gold.

Archimedes' screw, revolving inside a cylinder, drains water from a ship's hold.

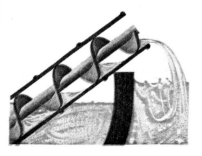

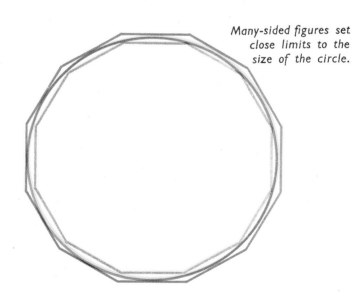

Many-sided figures set close limits to the size of the circle.

Archimedes chose two simple geometric figures to mark his tomb.

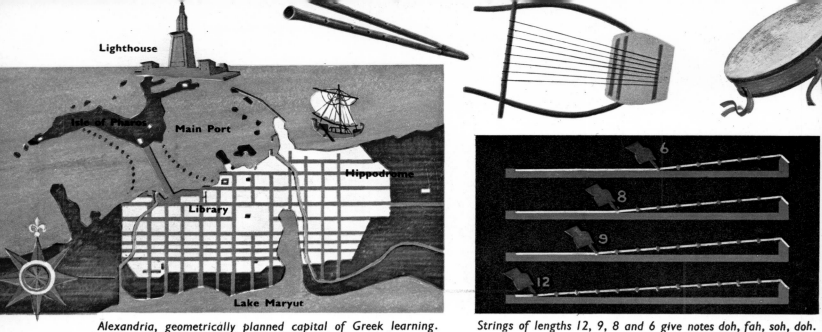

Alexandria, geometrically planned capital of Greek learning.

Strings of lengths 12, 9, 8 and 6 give notes doh, fah, soh, doh.

Hipparchus, who lived half a century after Archimedes, condensed the essentials of Greek geometry for the use of the astronomer and the surveyor in what we now call a table of sines.

We know that the angles of a triangle add up to two right-angles. If a triangle contains a right-angle and one known angle (A), the third angle, (B), must be the difference between a right-angle and A:

B is 90° minus A.

The ratio of the length of the side opposite A to that of the longest side is called the *sine* of A. This ratio is the same for all right-angled triangles which contain the same known angle (A). From the rule of Pythagoras, it is possible to find this ratio, or sine, when the known angle (A) is 60°, 45° or 30°. From another rule which he himself discovered, Hipparchus worked out many other sines, thus giving the surveyor or astronomer a wide range of angles with which to work.

The world of which Alexandria remained the capital of learning for seven hundred years was one which gave every encouragement to the development of arts and sciences which need the aid of mathematics. Sea-trade fostered the study of navigation and astronomy; frequent military campaigns called for more surveying and map-making; the demand for weapons of war led to a closer study of mining problems and mechanics. It is difficult to set a limit to the technical advances which might have been made if it had not been for the difficulty which the Greek world had in dealing with numbers.

The word arithmetic is a Greek word: but the Greeks did not mean by arithmetic what we ourselves mean—calculating with numerals. Perhaps it meant something closer to: getting fun out of figures. Numbers standing for the lengths of strings that would give the notes of the scale intrigued Pythagoras and his disciples. Figurate numbers, which stand for points that one can lay

In Greece music was a branch of mathematics.

ADDING ODD NUMBERS

The sum of the first two odd numbers is 2 × 2 = 4; the sum of the first three is 3 × 3 = 9; the sum of the first four is 4 × 4 = 16.

out in a geometrical pattern, had a special fascination. The best-known are the so-called triangular numbers, 1, 3, 6, 10, and so on, built up like this: 1, 1+2, 1+2+3, 1+2+3+4, etc. One of the sworn secrets of the Pythagorean Brotherhood was how to say what any particular number in the set is. The rule is simple: if asked to give the fifth number in the set, you multiply 5 by (5+1) and divide the result by two, which gives 15; if asked to give the twentieth number, you multiply 20 by (20+1) and divide the result by two, which gives 210.

Playing with pebbles may have given the Greeks the clue to a rule for finding the sum of consecutive odd numbers beginning with 1. If we add ten such numbers, the total is 10 × 10 = 100; if we add twenty, the total is 20 × 20 = 400.

ADDING CONSECUTIVE NUMBERS

1+2+3=6   1+2+3+4=10   1+2+3+4+5=15

$\frac{3\times(3+1)}{2}=6$ $\frac{4\times(4+1)}{2}=10$ $\frac{5\times(5+1)}{2}=15$

In numbers, the Greeks found mystery, magic and amusement.

Another kind of play with numbers and ideas is well shown in a conundrum put forward by Zeno, a very wise mathematician of Alexandria. Like all his friends, he knew very well that the swiftest runner in a race will overtake his fellows; but when he brought numbers into the argument, it seemed as though Achilles could not overtake a slow-moving tortoise given a good start.

The puzzle runs something like this: Achilles runs ten times as fast as the tortoise. He gives the tortoise a start of one-tenth of a mile. While Achilles runs this tenth of a mile, the tortoise moves on another hundredth of a mile; while Achilles covers that distance the tortoise goes another thousandth of a mile; and while Achilles covers it, the tortoise goes on another ten-thousandth of a mile. To the Greeks it seemed, by this argument, that Achilles should always be a little behind the tortoise. Yet they knew he would really overtake it. Why is the argument wrong?

In Egypt . . .

Mesopotamia . . .

and China, the earliest forms of writing consisted of a different picture to stand for each word.

There is no difficulty in answering the question from the practical point of view. With our own numeral system we can quite simply write the ordinary fraction 1/9 as the decimal fraction 0.i̇, in which the dot over the 1 is a short way of saying that 1 repeats itself over and over again, for ever. We thus see at once that $1/10 + 1/100 + 1/1000 + 1/10{,}000$, and so on for ever, is exactly 1/9. So Achilles will overtake the tortoise exactly one-ninth of a mile from his starting point. "For ever" does not mislead us. Our own number signs tell us we inevitably reach a limit however long we go on piling up such smaller and smaller fractions.

Strangely enough, the alphabet, which helped them in so many other ways, hindered the Greeks in the art of calculating.

The early priests of Egypt and Mesopotamia used pictures to stand for words or ideas. The man who wanted to read their inscriptions had to memorise thousands of separate picture signs. The Greeks, who inherited alphabetic writing, had only to master the shapes of a few letters and

Alphabetic writing, as on the Hebrew coins, began among the Jews and Phoenicians, and was later adapted and used by the Greeks.

Hebrew Coins

Greek Plaque

to memorise the sounds they represent. They hoped that the alphabet which so simplified word-language would also simplify number-language. So they began to use letters to stand for numbers.

At first they used the initial letters of words, just as we might use T for ten and H for hundred. The early Greek words for ten, hundred and thousand were deka (Δεκα), hekto (Hεκτο) and kilo (Χιλο), and they provided the number signs Δ for 10, H for 100, X for 1,000. To make up large numbers, the writer repeated these signs as needed. Apart from the fact that the early Greeks used new signs for 5, 50 and 500, their number script was much like that of Egypt. We can see this when we write the same number (3420) in both scripts.

𓏺𓏺𓏺 ℮℮℮℮ ∩∩
X X X   H H H H   Δ Δ

The later Greeks developed an entirely different style. They used the first nine letters of the alphabet to stand for the numbers 1 to 9, the next

The name inscribed on this Greek tomb is Democleides Demetrio.

Learning to spell: the word is the same as our flower-name, IRIS.

nine letters for 10 to 90, and the last nine for 100 to 900. They made any number a thousand times as large as its normal value simply by placing a stroke in front of it. The only advantage of this system was that large numbers took up less space. Thus 3420 was written as /ΓΥK. To offset this trifling convenience, however, the later Greek number-system made calculation more difficult. When an Egyptian priest wrote ℮℮℮ ∩∩ ||| (323) he could easily see that it meant three beads in the hundreds column of his abacus, two in the tens column, three in the units column. When a Greek wrote TKΓ (323) it told him nothing about how to lay out his abacus.

While the world was saddled with Greek number signs, there could be no simple table of multiplication such as we now learn when very young. The mathematicians of Alexandria did have scrolls of some tables to avoid doing all their calculations on the counting-frame, but a scroll would need to be almost endless to hold all the numbers we need to multiply in modern astronomy.

Today we use letters as a kind of shorthand in mathematics. Instead of saying that the area of a triangle is equal to its base multiplied by its height and divided by two, we say it is $\frac{bh}{2}$. But for the Greeks, who used every letter of the alphabet to stand for a different number, such a shorthand was almost impossible.

The Romans, who succeeded the Greeks as masters of the Mediterranean world, patterned

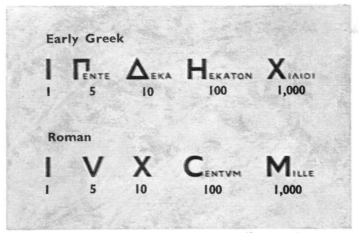

Roman numerals followed the pattern set by the early Greeks.

their number-system on that of the early Greeks. We rightly think of the Romans as a nation of conquerors, but they were never able to conquer the art of calculating as we know it today. Even simple multiplication was a slow and space-consuming process. The Roman merchant might use numerals for recording, but calculating was still a task for the slave working with an abacus.

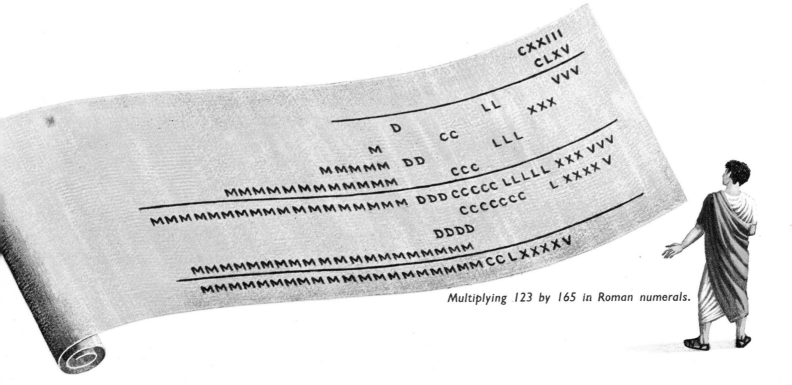

Multiplying 123 by 165 in Roman numerals.

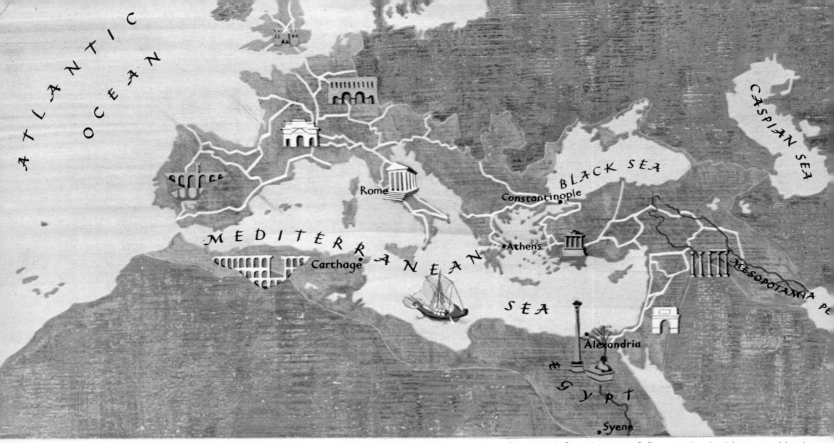

Wherever the conquering Roman legions went, they brought law and order and left a legacy of fine roads, buildings and bridges.

About two thousand years ago, the Roman legions conquered the whole of southern Europe, all Gaul, most of Britain, the northern fringe of Africa and a large area of west Asia. Throughout the conquered regions they built a vast network of good roads and fine bridges. They built also a new system of law and government. Yet the Greek language long remained the language of learning.

In the fourth century A.D., the Empire was divided into two parts; the western half with its capital at Rome and the eastern with its capital at Constantinople, now Istanbul. By then the Empire was under strong pressure from the revolt of non-Roman troops inside its borders and from powerful tribes warring on its frontiers. Gradually the west lost contact with the Eastern Empire which still preserved the language and tradition of Greek learning.

When the mighty Roman Empire decayed, a small remnant of the eastern half, embracing Greece and the country a little north of it, remained intact, keeping a little of its old glory. Meanwhile the west soon forgot utterly the language and science of Greece. What learning remained there was preserved by monks who brought the art of writing to Gaul and Britain.

CAESAR
DE BELLO GALLICO
LIBER PRIMUS.

1. GALLIA est omnis divisa in partes tres, quarum unam incolunt Belgae, aliam Aquitani, tertiam qui ipsorum lingua Celtae, nostra Galli appellantur. Hi omnes lingua, institutis, legibus inter se differunt. Gallos ab Aquitanis Garumna flumen, a Belgis Matrona et Sequana dividit. Horum omnium fortissimi sunt Belgae, propterea ... cultu atque humanitate provinciae ... nimeque ad eos merc...

Gaul is divided into three parts, viz. Belgium, Gaul proper, and Aquitania.

Roman numerals, Roman engineering and the Latin language have remained part of our western heritage to the present day.

Because Western Europe learned its letters from monks of the Roman Church, Latin, the language of the Church service, there became the language of learning. Even in the time of Columbus books about divinity, law and medicine were mostly in Latin. The only numerals used were Roman numerals which label each groove of the abacus differently, as M, C, X, I. Today we still see traces of this long Latin influence. We still label our books as Vol. I and Vol. II; and the dials of our clocks often carry Roman numerals.

Before the West could make any real progress in the art of calculation and in science, it had to have help. Help came from the East.

A Roman aqueduct in Turkey.

Numbers and Nothing

ONE OF THE world's oldest civilisations grew up in the valley of the River Indus, in India. Like those of the Nile and the Euphrates, it learned its first lessons in mathematics through astronomy, the gateway to time-reckoning and to temple-building. Several centuries before Rome rose to power, the mathematicians of India had found a close value for π. In the arithmetic of trade, the merchants of India were the equals of those of Mesopotamia.

Until about two thousand years ago, they probably used numerals made up of horizontal strokes. But when they began to use dried palm leaves as writing material and developed a flowing style of writing, they also began to join up these strokes, so that = became ⇁ and ☰ became ⇅. In this way they gradually built up different signs for each number up to nine. Each sign could conveniently be used to indicate the number of pebbles in *any* groove of the abacus.

Had progress stopped short there, it would not have amounted to much. If ⇁⇁ merely stands for two pebbles in any two grooves, it can have many different meanings, such as twenty-two, two hundred and two, two thousand and twenty, and so on. We need to be told not only how many pebbles in a groove, but also which groove they are in.

Somewhere in India, some unknown person, probably a counting-house clerk, hit on a device which does this for us. He used the figure on the extreme right to stand for pebbles in the units groove, the next figure to the left to stand for pebbles in the tens groove, and so on. To indicate an empty column, he used a dot, just as we now use a zero. Thus ⇁⇁ could mean only 22. ⇁.⇁. could mean only 2020.

With this system, we no longer rely on space-consuming repetition; and we can record the same number on any groove of the abacus by using the same sign. But saving space is only a small advantage, as the later Greeks must have learned. The great advantage of the Hindu system is that it enables us to *calculate* with numerals.

The ancient systems of writing – Egyptian, Babylonian, Greek, Roman and Chinese – all relied on the use of different symbols for the same number of pebbles in different grooves of the abacus. Before you could do written or mental calculations with them, you would therefore need to learn a different table of addition and of multiplication for each groove. When you have only nine different

revolution in the art of calculation. The mathematicians of India began to think of fractions and to write them in the way that we do. By 500 A.D. India had produced mathematicians who solved problems which had baffled the greatest scholars of antiquity. The mathematician, Varahamihira, was able to calculate how to forecast the positions of planets; Aryabhata stated a rule for finding square roots and gave a value for π which is still good enough for most purposes today – 3.1416.

By about 800 A.D., Indian traders, following the age-old caravan route which passed through Persia into Mesopotamia, brought news of the new numerals to Baghdad which was then rapidly becoming the world's greatest city of learning.

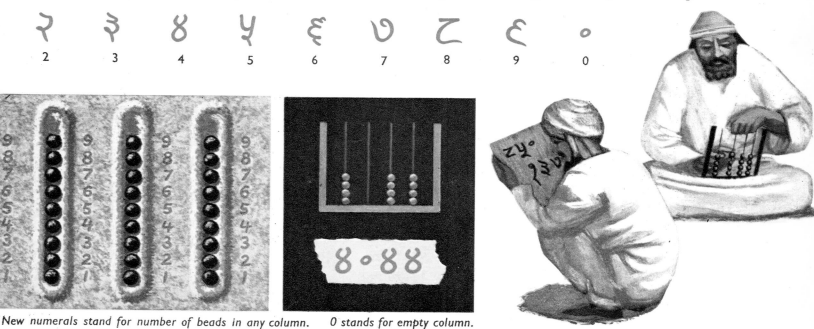

New numerals stand for number of beads in any column.   0 stands for empty column.

signs, each of which can show the number of pebbles in any groove, and a zero to indicate empty grooves, you need learn only one simple table, once and for all. You can carry over in your head because there is only one simple table to remember. The Hindu number-language quickly led to a

Palm-leaf manuscripts.

As caravans of Hindu traders moved westward along the age-old trade routes, they carried not only merchandise but also ideas.

To Baghdad, magnificent capital city of the great Moslem Empire, came Hindu merchants with their wonderful new numeral system.

Early in the seventh century, Mahommed, founder and prophet of the Moslem religion, united the whole of Arabia under his leadership. For more than three centuries after his death, his followers carried the new religion across the whole of North Africa, into Spain and Portugal, and eastward through Asia beyond the River Indus.

About 762 they founded the city of Baghdad and made it the seat of government of a rapidly-growing empire. About forty years later, under the Caliph, Harun ar-Rashid, it became the capital of learning of the western world, just as Alexandria had been during Greek and Roman times.

In learning, Baghdad made the best of both worlds, East and West. Merchants and mathematicians from the East brought with them the new number signs and the arithmetic of India. Heretics, who had fled from the West, brought copies of scientific works written while Alexandria was still at its prime. These included treatises on astronomy and geography and Euclid's geometry. By order of the Caliph, Moslem scholars translated such works into Arabic, the language of their sacred book, the Koran. Thus, the science and geometry of Greece became available throughout the Moslem world, now equipped with an arithmetic far better than the best the Greeks had known.

To this growing body of knowledge the East made two other contributions. Chinese prisoners, captured during a skirmish on the frontier, taught

Chinese brought paper-making.

Persians brought astrology.

Baghdad the art of paper-making, while Persian astrologers, who added a spice of eastern magic to a sound knowledge of the heavens, gave the Caliph's court a keen interest in astronomy.

In observatories built by command of the Caliph, astronomers advanced the science of map-making far beyond the level it had reached in Alexandria. In the schools of Baghdad trigonometry flourished. Because they had mastered the new arithmetic of India, Moslem mathematicians could make much fuller use of the geometry of Euclid and of Archimedes. The astronomer equipped the mariner with nautical almanacs for navigation by sun and stars and gave him improved instruments, designed in the observatories. The geographer had new and better tools for land-survey.

Never before in history had knowledge advanced in a single century as it did between 800 and 900 A.D., where East met West in Baghdad.

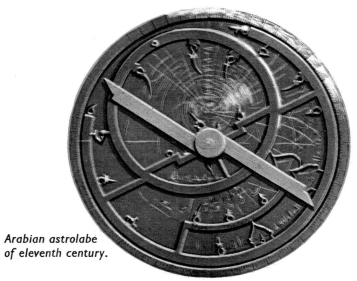

Arabian astrolabe of eleventh century.

Crown copyright. Science Museum, London.

Refugees from the west brought Greek geometry into Baghdad.

In elaborate new observatories, Arab astronomers accumulated an ever-increasing knowledge of the movements of the heavenly bodies.

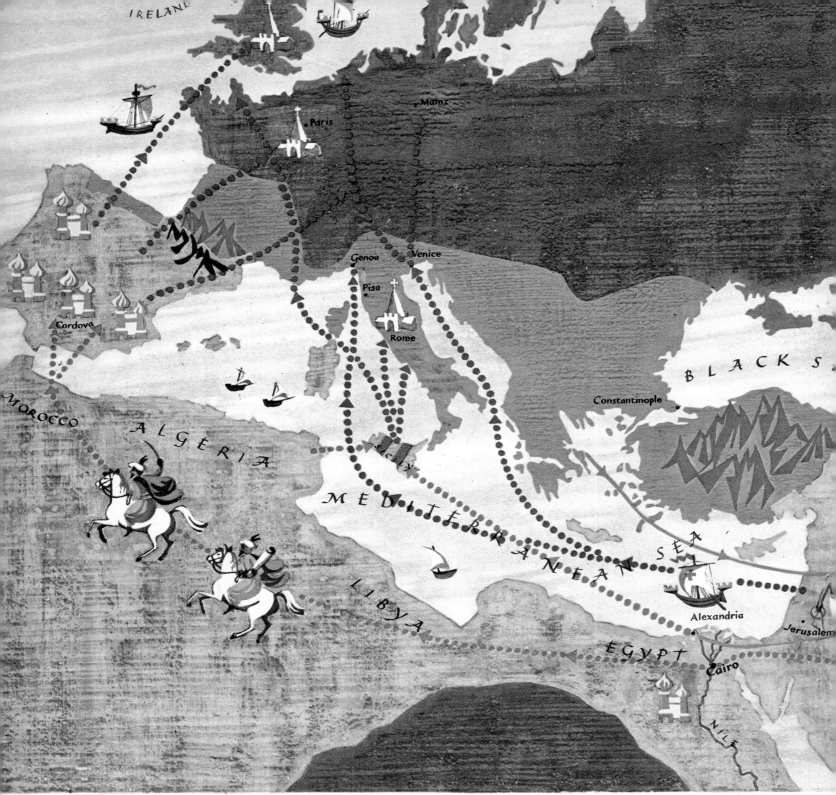

How various kinds of knowledge mingled in Baghdad and slowly penetrated into Europe.

Our map shows how the new learning swept across the western world. The drawings near the orange arrows show what kinds of knowledge met and mingled in Baghdad. Pale green areas show where the conquering Moslems carried knowledge with them. Red arrows show the routes along which learning spread beyond the Moslem Empire.

By the year 1000 A.D., the greater part of the old Roman Empire had come under Moslem rule. Spain, now occupied by Moorish followers of Mohammed, especially enjoyed the benefits of the new learning. In its Moslem universities, students could study the geometry of Greece, the arithmetic of India, and the sciences of astronomy, of trigonometry and of geography which the scholars of Baghdad had done so much to advance.

Early in the twelfth century a Christian monk, Adelard of Bath, disguised himself as a Moslem and studied for many years at the University of Cordova. There he made translations of the works of Euclid and of the Moslem mathematician, Alkarismi. These he smuggled back to Britain.

During the following century, Jewish physicians, trained in Spanish universities, carried the new learning to France and to Italy, founding medical schools in Christian universities which had formerly ignored the teaching of science. Moslem culture also reached Europe along the sea-lanes from Sicily and by the routes crusaders followed when returning from the Holy Land.

By 1400 A.D. the merchants of Italy, France, Germany and Britain were using the new numerals, and schools for the teaching of the new arithmetic began to spring up throughout Europe. Half a century later, when printing began, textbooks of arithmetic and nautical almanacs were among the chief products of the printing press.

In 15th-century Portugal, the sailor prince, Henry the Navigator, set up a school where Jewish teachers, trained in the universities of Spain, instructed pilots in navigation. The success of Columbus is in no small measure due to the pilots he enlisted: Jewish experts trained in Moslem astronomy and Moslem mathematics.

HINDU, 800 A.D.

ARABIC, 900 A.D. ONWARDS

SPANISH, 1000 A.D.

ITALIAN, 1400 A.D.

237 × 146 = 34,602

934 × 314 = 293,276

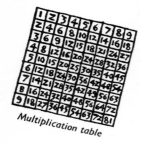
Multiplication table

From drawing in a 15th-century arithmetic.

As the new numerals passed from one land to another, they frequently changed their shape, but they always kept to the pattern of nine number-signs and a zero. It would be wrong to imagine, however, that they were immediately and everywhere acceptable. In the thirteenth century a law forbade the bankers of Florence to use them, and a hundred years later the University of Padua insisted that price-lists of books should appear in Roman numerals. But by the fifteenth century, Hindu numerals were in general use for navigation and commerce throughout Western Europe. For centuries after that many people still used the abacus and counting board, but more and more were eager to learn the new arithmetic.

Textbooks which poured from the early printing-presses spread far afield convenient ways of writing down tables and of setting out problems in addition, subtraction, multiplication and division. Many of these methods, like those shown on this page, are now interesting only as museum pieces. Much more important is the fact that the early textbooks introduced new shorthand signs into arithmetic, such as + and − . These were not deliberate mathematical inventions. It is probable that they were originally warehouse signs, used to indicate which packages were overweight, which underweight. As these early signs proved their usefulness, others were gradually introduced: × (multiply), ÷ (divide), ∴ (therefore), = (equals).

Long before modern numerals reached Christian Europe, Hindu and Moslem mathematicians had

Arithmetic may have borrowed the signs + and − from the merchants' marks for overweight and underweight.

discovered many tricks for solving number problems of the kind we now deal with by algebra. In fact, algebra is an Arabic word, but it would be a mistake to think that the Moslem mathematicians taught algebra as we now learn it. Although they no longer used letters to stand for numbers, they had never hit on the idea of using them to state a rule or problem involving numbers in a short, snappy way. The only shorthand sign they used was $\sqrt{}$ for square root.

It was not until about 1600 that algebra, as we know it, had gradually taken shape. We can see how the new system developed by working out a simple problem: if the result of multiplying a number by two and dividing the product by three is forty, what is the number? The Hindu and Moslem mathematicians might explain the solution in some such words as these: Since two-thirds of the number is forty, one-third of it is half forty, which is twenty; and the number itself is three times this, which is sixty.

An early arithmetic might have put it this way: Find the number, if $(2 \times \text{number}) \div 3 = 40$. The solution would be written something like this:

$$\frac{2 \times \text{num.}}{3} = 40; \quad \frac{\text{num.}}{3} = \tfrac{1}{2}(40) = 20; \quad \text{num.} = 3 \times 20 = 60$$

In modern algebra we would shorten the word number to n, drop the sign $\times$, and write the solution in simple orderly steps, as shown below.

The Moslem teacher of 1200 A.D. could certainly have given a rule for solving any problem of this sort, but he would have given it in a long, cumbersome way, like this: If you know what the answer is when you multiply any number by a second and divide it by a third, you can find the number itself by multiplying the answer by the third and dividing the result by the second.

Today we would write this more snappily by putting n for any number, s for the second, t for the third, and a for the answer. The rule then becomes much easier to remember:

$$\text{If } \frac{sn}{t} = a, \quad n = \frac{ta}{s}$$

With new numerals, a new arithmetic and the beginnings of the new algebra, Europe was in a good position to tackle the practical problems that faced it in the Age of Discovery, soon to begin.

Bead-calculation was used long after written arithmetic began.

In the new schools of Europe, men worked to improve methods of calculation.

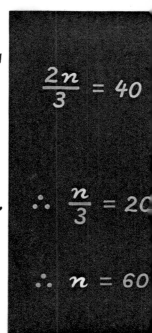
From words to the shorthand of algebra.

Graphs and Gravity

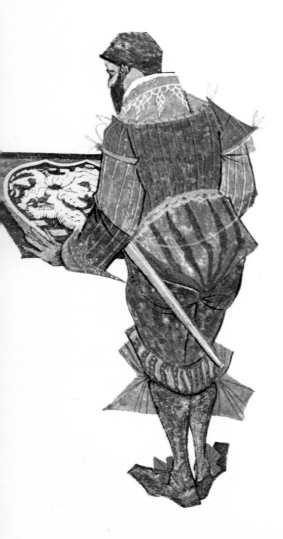

Columbus and Drake, Vespucci and Magellan, and all the other great sea-captains whose ships opened up new east-west sea-lanes across the Atlantic, were faced with one problem that sometimes proved a matter of life or death – the problem of finding longitude. Long after the days of these pioneers, one captain, when almost within reach of the island for which he was bound, imagined that he had sailed too far west and passed his objective. He then sailed east for three hundred miles before realising his mistake and turning westward once more. During the wasted voyage of six hundred miles many of his crew died of hunger or scurvy.

Before there were new and easier methods of finding longitude, the explorer-mariner had no means of locating the positions of ports accurately on his map. To fix your longitude, you need to know the time where you are and to compare it with the time at some other fixed point. By comparing the times when eclipses of the moon were visible at different places, the geographer Ptolemy of Alexandria was able to fix, roughly, the longitude of some half-dozen places. Moslem astronomers and geographers knew the longitude of perhaps a score of towns. This was a great gain; but the captain of an ocean-going ship needs a way of fixing the longitude of any place at any time, and for this he must have dependable and accurate time-keeping instruments.

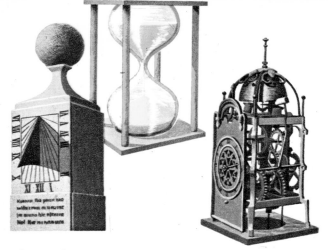

To early time-keepers, the Middle Ages added only crude clocks.

In the Age of Discovery, with ships crossing the world, men therefore needed to measure minutes and seconds accurately. For this purpose crude weight-driven clocks such as churches and monasteries had installed during the four previous centuries were of no more use than the candle-clocks, the sun-dials and the hour-glasses on which the ancient world depended.

The first clue to accurate measurement of small intervals of time was discovered in 1583, when Galileo, a young Italian medical student, watched a lamp swinging to and fro in Pisa Cathedral. Timing its motion by the beat of his

A swinging altar-lamp gave a clue to the law of the pendulum.

pulse, Galileo found that all swings, whether wide or narrow, took the same time.

Later on, when Galileo gave up the study of medicine to take up mathematics and physics, he used a home-made water-clock to check the accuracy of this observation. While a pendulum was swinging, he allowed water to flow from a hole at the bottom of a large vessel and fall into a small one below it. If the weight of water that escaped during two separate swings was the same, he knew that both had taken an equal time.

His experiments showed that the time of swing depends only on the length of the pendulum. To double the time of swing you must make your pendulum four times as long; to treble the time of swing you must make your pendulum nine times as long. The length of the pendulum varies in the same ratio as the square of the time of swing. We now know that this rule, while correct for narrow swings, is not quite accurate when the pendulum swings through a very wide arc. In 1657 the Dutch scientist, Huygens, made use of Galileo's discovery to produce accurate pendulum-regulated clocks.

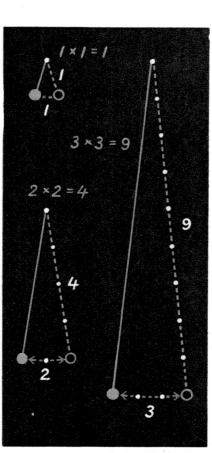

Red lines show length, blue show time. Length determines the time of swing.

Early pendulum clock.

As a ball rolled down a slope, Galileo measured its acceleration with a water-device.

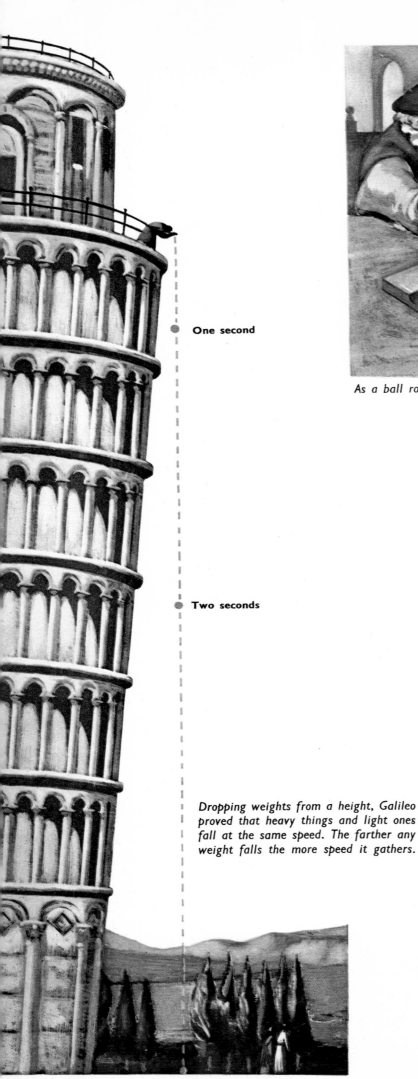

- One second
- Two seconds

Dropping weights from a height, Galileo proved that heavy things and light ones fall at the same speed. The farther any weight falls the more speed it gathers.

- Three seconds

Before Galileo's time, people believed that the heavier an object was, the faster it would fall. Galileo's pendulum experiments, however, disproved this. For he found that the weight of the bob at the bottom of the pendulum has no effect on the time of swing. To settle the matter beyond dispute, he dropped two different weights simultaneously from the Leaning Tower of Pisa. Both the heavy one and the light one hit the ground at the same instant.

Galileo recognised that both weights increased their speed, or accelerated, as they fell. In his day there were no stop-watches for split-second timing. So he found it impossible to measure the acceleration directly. He realised, however, that gravity acts on a ball rolling down a slope just as it acts on a falling weight; but the slope itself then slows down the speed of the ball. He therefore

Sixteenth-century effort to apply mathematics to range-finding.

The cannon completely revolutionised warfare. Once-impregnable forts, high up on hilltops, made easy targets for the new weapon.

rolled a ball down a sloping board and timed it as he had timed the swing of the pendulum.

He found that in two seconds the ball rolls four times as far as in one second; in three seconds it rolls nine times as far as in one second. The distance it rolls varies in the same ratio as the square of the time it rolls.

This discovery makes it possible to work out the kind of path a cannon-ball follows as it hurtles through the air. At the instant it leaves the mouth of the cannon, it would move in a straight line pointing in the same direction as the gun-barrel, if there were no force of gravity to pull it downwards at a uniformly increasing rate. But because of the pull of gravity, it travels along the kind of curve which we call a parabola.

Before the time of Galileo, mathematicians had tried, without much success, to advise the artilleryman about how to decide the correct elevation for the cannon when he knew the distance of the target. When it was possible to understand how gravity affects the flight of the cannon-ball, it was also possible to work out tables of elevation, based on the distance of the target. This distance, together with the speed of the ball, decides how long the ball will be in flight and hence how long the force of gravity will be acting on it.

In the seventeenth century, military engineers trained in mathematics designed new fortifications to withstand attack by cannon. Low-built forts protected by earthworks replaced hillside fortresses which enabled the defenders of earlier times to fire down on their attackers. The new ones confronted the attackers with a more difficult target, while the defenders, with cannons placed low, could answer their fire as effectively as from a height.

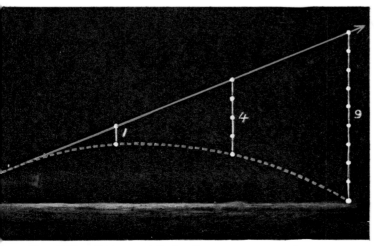

Shot falling faster and faster from course, travels in a curve.

Knowledge of cannon-ball's path changed system of fortification.

The problem of finding longitude also ties up with the movement of a ball: the great ball we call our Earth. Each day the earth makes one complete spin round on its axis, from west to east. All the time, part of it is turning out of the sunlight into the shadow and part is turning out of the shadow into the sunlight. When it is noon at any given place, it is later than noon to the east and earlier than noon to the west.

Geographers divide the earth into 360° of longitude (180° for the eastern hemisphere and 180° for the western). Since there are 24 hours, or 1,440 minutes, in a day, the difference in time for each degree is 4 minutes (1440 ÷ 360). So, if we know our local time and the time at some other place at any particular moment, we can work out the difference in longitude. If it is 12 noon in London and 7 a.m. where we are standing, our local time is 300 minutes earlier than that of London; we are therefore 300 ÷ 4, or 75°, west of London – roughly the longitude of New York.

Columbus threw wood or barrel overboard from bow.

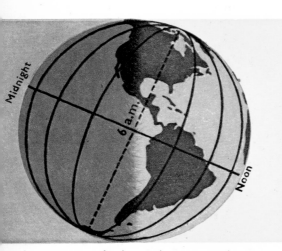 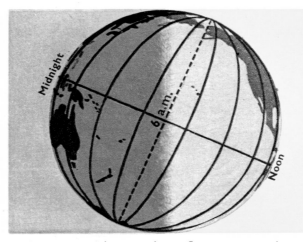 

A sextant.

As the earth spins round on its axis, time moves on evenly everywhere. If, at any particular moment, a seaman knows his local time and the time at any other fixed point, he can find his longitude.

In the age of Columbus, a ship's captain could make a very close approximation to his correct local time with the help of an astrolabe, but he had no convenient method of finding the time at another fixed point. For this, he would rely on his almanac, which told him the time at which an eclipse of the moon or the disappearance of a planet behind the moon's disc might be visible at his home port. He had then to wait until he saw such an occurrence, to take his local time at that moment, and to compare it with the time shown in the almanac. His almanac would always give him the home-time of an eclipse of the moon or the home-time at which the moon would "occult"

Almanac gave the time of an eclipse at home port.

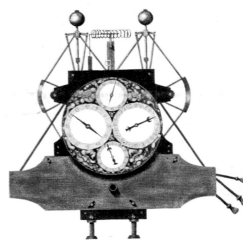

Harrison's first chronometer

The time his ship took to pass it gave him his speed. His compass gave his approximate direction.

a planet, or hide it from view; but eclipses of the moon and occultations by planets do not occur often in the course of a year.

Between times, captains had to experiment with their own ways of keeping check on their whereabouts. Having set his direction in a rough and ready way from the compass, Columbus used to throw a piece of wood or a barrel overboard from the bows. If his 50-foot ship took ten seconds to pass it, he knew he was travelling at 300 feet a minute, or roughly three-and-a-half miles an hour.

The longitude problem was solved completely only after the middle of the eighteenth century, when ships were first fitted with sextants and chronometers. The sextant gave the navigator a more accurate means of finding local time, and the chronometer enabled him to carry the time of his home port wherever he went. The first chronometer, i.e. a clock which keeps accurate time over a long sea-voyage, was invented by a self-taught English carpenter, John Harrison, at the time when Benjamin Franklin was making his great discoveries about electricity.

A century later, all sea-faring nations agreed to set their chronometers by Greenwich standard time and to measure longitude from the line on which Greenwich Observatory, London, stands.

Astrolabe gave local time.

Greenwich Observatory about 1700

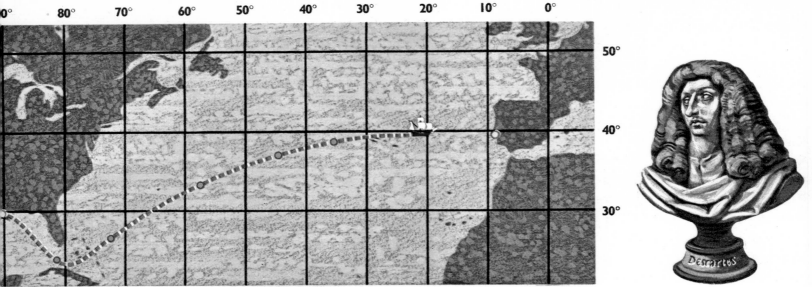

On a map marked with lines of latitude and longitude, one curve can sum up a ship's voyage.

During the sixteenth and seventeenth centuries, navigators began to plot the day-to-day position of their ships on maps marked with lines of latitude and longitude. A connecting line drawn through all these points gave the navigator a convenient summary of the voyage.

Mathematicians were already trying out much the same technique to represent figures by paths which a moving point traces on the sort of grid we now call a graph. If we make such a grid with vertical lines to show time and horizontal lines to show distance, we can easily plot Achilles' race with the tortoise. One line shows where Achilles starts and how fast he runs; another shows where the tortoise starts and how fast it runs. The point where the lines cross shows where Achilles overtakes the tortoise.

The first man to realise clearly how useful graphs can be was René Descartes, a great French mathematician of the seventeenth century. As a simple example of one use to which we can put them you may try solving the problem of finding x when we are told that $4x^2 - 4x - 12 = 3$.

A statement of this kind tells us that one quantity equals another, so we call it an equation.

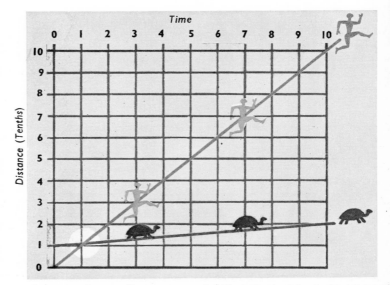

On a graph, two lines sum up Achilles' race with the tortoise.

As a first step to getting the answer, it is usual to make one side of the equation zero, but we can *keep* both sides equal only if we take the same quantity from each. We therefore take 3 from each side and re-write it as $4x^2 - 4x - 15 = 0$. We now make guesses about the value of x and draw up a table to show what the right-hand side of the equation will be if each guess is correct. (In working out our table we need to remember that multiplying like signs gives a plus, multiplying unlike signs gives a minus.)

If we choose whole numbers for the values of x in our table, we find that none of them yields the right result; that is to say 0 on the right-hand side. But if we make a graph by plotting all our results as points placed vertically for successive values of x spaced equidistant horizontally, we can get a curve like a ship's course by joining each

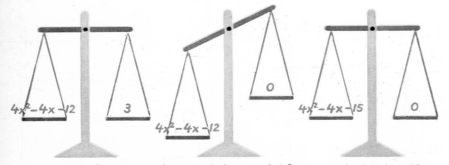

An equation keeps in balance only if we treat both sides alike.

point. This curve cuts the zero line at two points. $2\frac{1}{2}$ and $-1\frac{1}{2}$. We thus see that there are two values of x which will solve our equation correctly, as is always true of this kind of equation, which we call a quadratic.

Descartes was also one of the first mathematicians to write out equations with letters and signs which we use today; but most of all, he made a closer tie between geometry and algebra than ever before. He first used algebra to state rules for drawing certain geometrical figures.

The Greeks had studied only curves one can draw with the help of compass and ruler. Descartes contended that any curve is worthy of study if we can state a rule for drawing it. We cannot draw the curve called the parabola which corresponds to the flight of the cannon-ball, if we stick to the compass and ruler recipe, but it was important for the mathematician to study such a curve in an age when the cannon could decide the fate of a nation. By the use of the graph, Descartes was able to state a rule which does enable us to draw a parabola.

The scientists of the period were becoming increasingly aware of the importance of another curve, the ellipse. With a few exceptions, such as Aristarchus and Philolaus, Greek astronomers believed that the sun moves round the earth, and

Value of x	Value of $4x^2 - 4x - 15$	Result
-2	$16 + 8 - 15$	9
-1	$4 + 4 - 15$	-7
0	$0 - 0 - 15$	-15
1	$4 - 4 - 15$	-15
2	$16 - 8 - 15$	-7
3	$36 - 12 - 15$	9

Points where curve cuts zero-line give correct values for x.

until about 1540 Western Europe accepted that belief. Then Copernicus, the great Polish astronomer, revived the theory of Aristarchus, that the earth and the planets circle round the sun. During the next hundred years this theory was confirmed in broad outline by other astronomers, including Tycho Brahe, Kepler and Galileo, but Kepler discovered that the track of a planet round the sun is not exactly a circle. It is an ellipse—the figure which we now draw by moving a tightly-stretched loop of cord around two fixed pins or pegs.

In an age when accurate navigation depended more than ever on the work of the astronomer, the new geometry of Descartes made it possible to state as an algebraic formula the rule which enables us to draw an ellipse on a graph.

Loop drawn round two pegs produces an ellipse.

Uraniborg, Danish for 'Tower of Heaven', where Tycho Brahe observed heavens.

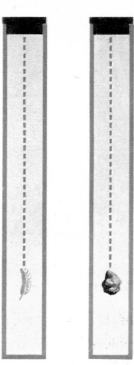

In the early air-pump, a vacuum was created by pumping water from a sealed vessel. In a vacuum, feather and stone fall at same speed.

Isaac Newton, the greatest scientist and mathematician of the Age of Discovery, gathered the threads of observation and reason, spun by so many earlier scientists, and wove them into a satisfying pattern.

From earliest times, men have studied the motions of sun, moon and stars, but Newton was the first to give a satisfactory theory of their movements. Kepler who discovered that the planets move in ellipses round the sun, could never understand why they do so. Galileo, who understood how the force of gravity explains the path of a cannon-ball, did not realise that the same force might explain the path of the planets.

Before Newton gave his explanation, an important invention had advanced knowledge beyond the level of Galileo's time. The air-pump had made it possible to experiment with falling bodies in a vacuum, and so to obtain more accurate information about gravitation. Reasoning from how we trace the flight of the cannon-ball from what we know about the behaviour of falling bodies, Descartes stated the rule that any moving body will continue to move in a straight line unless some force halts it or changes its direction.

Thus Newton's problem was not to explain why the planets keep moving. The question he tackled was why they move in a closed curve rather than in a straight line. His solution was that the force of gravity throughout the universe

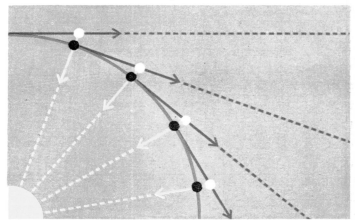

Sun's mass pulls planet steadily from straight path it would otherwise follow, so that planet travels along closed curve.

acts in accordance with the same laws as on our own earth. Just as the mass of the earth pulls a weight towards its central point, the mass of the sun pulls a planet towards *its* central point. In the absence of gravity, a planet, like a cannon-ball, would travel in a straight line; but the pull of the sun moves it away from that line. Newton demonstrated how the speed of the planets and the pull of the sun together keep the planets in the closed curve which they follow.

One thing which contributed to the tremendous progress in astronomy in the days of Galileo and Newton was the invention of the telescope. It seems that the first telescope was made in 1608 by a Dutch spectacle-maker named Hans Lippershey, but it was Galileo who first used a telescope

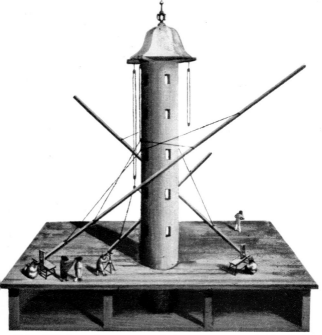

This 150-ft. telescope needed a tower to support it.

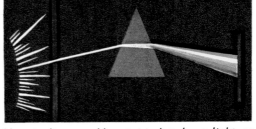

Newton discovered how prism breaks up light, and made small reflecting telescope which gave clearer image.

of his own design to study the heavens. The weakness of the early telescope was that it gave a blurred image of the object on which it was focused. Some astronomers tried to remedy this by altering the shape or position of the lenses, others by lengthening the whole instrument.

While investigating the properties of light, Newton discovered the varied coloration which results when a beam of sunlight passes through a glass prism. He realised that the light passing through the lens of Galileo's telescope behaved in the same way, thus blurring the image. He therefore designed a new telescope in which light from the object was reflected from a curved mirror on to a flat one and thence to the eye-piece, without having to pass through a lens at all.

In the days of Newton, scientific academies were being founded throughout Europe, and, more than ever before, scientists of many lands were pooling their knowledge. Thus it happened that two outstanding men, both drawing from the common pool, made the same great advance in mathematics, independently of each other and at the same time. Leibniz in Germany and Newton in England both founded a new and fruitful means of calculation called the infinitesimal calculus. It has revolutionised every branch of science which plays a part in modern industry.

Scientific academy of Newton's time

Power and Precision

FOR THOUSANDS of years, man has harnessed the wind to drive his sailing ships. For hundreds of years he has used the power of the wind and of fast-flowing streams to turn mill-sails and mill-wheels. Yet right up to the time of Newton and Leibniz most of the world's work – the lifting and carrying, hewing and hammering, making and mending – was still done by muscle-power. By then the need for new sources of power was becoming urgent. The miners of western Europe, and especially those of Britain, were sinking deeper shafts than ever before. Muscle-powered pumps could no longer cope with the large quantities of water which accumulated in the pits.

By the close of the seventeenth century, Denys Papin, a Frenchman, and Thomas Savery, an Englishman, had both succeeded in making crude pumps driven by steam. Within a few years Thomas Newcomen made the first steam-powered piston engine. Fifty years later James Watt fitted the steam-engine with a separate condenser, which cut down waste of heat and fuel. He also invented a means by which the steam-engine could be made to turn wheels.

During the century that followed Watt's inventions, steam-power rapidly changed the whole way of life in the western world. Industry moved away from the country cottage into the factories of huge industrial cities which sprang up near coalfields, where fuel for steam-engines was cheap and plentiful. Smoking funnels replaced white sails along the world's sea-routes. The clip-clop of the coach horse died out on the highway and made way for the rattle of steam locomotives carrying freight and passengers along the new railroads.

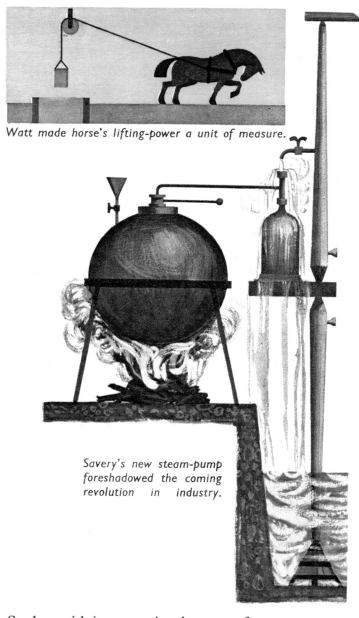

Watt made horse's lifting-power a unit of measure.

Savery's new steam-pump foreshadowed the coming revolution in industry.

to find. New kinds of measurement are more easy to understand if based on older ones we already use. When improved oil lamps and gas lamps were taking the place of candles at the beginning of the new industrial age, the illumination they gave was at first measured in candle-power.

In the time of Watt, all steam-engines worked at much the same pressure. It was thus possible to estimate the horse-power of an engine from the size of its cylinder. As design became more varied, indicators or steam gauges came into use to measure the pressure of steam generated in the cylinder in pounds per square inch.

Many units of measurement we use today would have puzzled the engineers and scientists of Watt's time. When we speak of volts and amperes in connection with electricity, or therms and calories in connection with heat, we are using a language of precise measurement devised to meet the needs of the age of power.

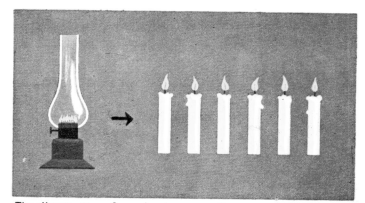

The illumination of new lamps was at first measured by the unit of illumination of the candles, which they were to replace.

Such rapid increase in the use of steam power was at first largely due to the way in which Watt and his business partner, Boulton, were able to convince customers of the usefulness and cheapness of the engines they made. They found by experiment that a strong horse can raise a 150-lb. weight, suspended over a pulley, 220 feet in one minute. If one of their engines could raise ten times that weight through the same distance in one minute, they classed it as a ten horse-power model. The customer could then compare the cost of buying fuel for such an engine with the cost of providing keep for ten horses, and usually he found that it would pay him, in the long run, to lease the engine.

It may seem strange that horse-power became a standard unit of power-measurement just at the moment when horses were losing their importance in industry, but the reason is not hard

Steam-gauges measured pressure in pound-per-square-inch units.

At night large towns are ablaze with light, near dawn almost in darkness. Power stations must anticipate change in power-demand.

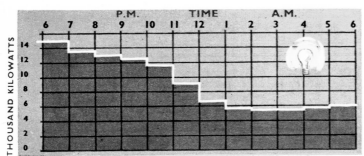

Each bar shows the power consumed in one particular hour.

In setting the world on the road to greater power, Boulton and Watt also ushered in the age of large-scale production. Boulton once wrote: It would not be worth my while to make for three counties only; but I find it well worth my while to make for all the world. The manufacturer who produces goods or services on such a scale cannot long remain content with accounts which show only his income and expenditure, profit or loss. He must plan production ahead, and to do that he needs to know the answers to a host of questions: Does the demand for his products change from season to season? Where do his goods sell best? Where and how can he improve the sales of his goods?

A record of information which may help to answer such questions is often kept in the form of simple diagrams. For example, a power station may record, by means of a bar-chart or histogram, how much electricity it supplies in a day. The height of each bar of such a chart indicates the amount of power used in one particular hour. An exporter may prepare a pie-chart, in which the

Pie-chart shows where sugar from Cuba was sold in a recent year.

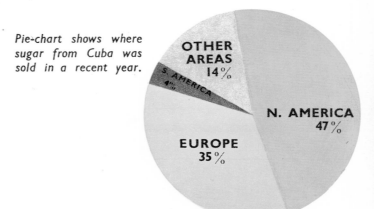

Loading sugar-cane for export.

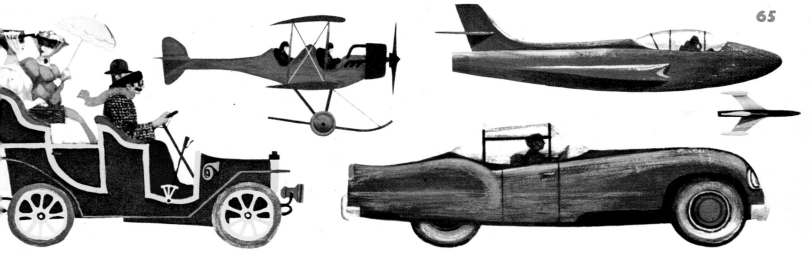

The great change in the design of automobiles and airplanes is the result not of whim but of research and of applied mathematics.

whole area of a circle represents his total overseas sales, and the areas of various segments, or slices, represent the sales to particular regions.

Progress in accountancy is only one feature of the age of power. Perhaps a more important one is progress in design.

When we compare the shapes of automobiles or of airplanes of forty years ago with those of today we can see how great such progress has been. We may or may not prefer the new look for its own sake; but it spells greater efficiency. The streamlined design enables the modern machine to move smoothly and rapidly with a minimum expenditure of power.

Change of design has not come about by the whim of fashion. It relies on the research work of the engineer whose calculations rely on the work of the mathematician. A recent pamphlet on aerodynamics, the science which concerns itself with wind forces, wind speeds and streamlining, says: Advanced mathematical treatment, continually checked and modified by experimental research, is now an essential tool.

Thus we see that mathematics is just as closely bound up with the problems of real life today as it was when the priests of Egypt planned the great pyramids. It may need the practised eye of the aircraft-designer to detect the mathematics that lie behind the subtle curves of a modern jet-plane, but there are other modern structures which proclaim their mathematical origins as clearly as do the pyramids. When we look at the blue-print for a suspension bridge, we are clearly seeing just the kind of graph-line that Descartes might have drawn, and we recognise the finished bridge as a graph in steel.

The real-life problems of a fast-moving world are far more complex than those of ancient Egypt, when the shortest unit of time was for most purposes the hour. As the problems which crop up in the world's work have come to be more complicated, mathematics has come to be more complicated in the effort to solve them. Fortunately the mathematician of our own time has at his command aids to rapid calculation such as his predecessors never dreamed of.

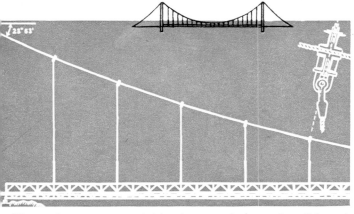

The modern suspension bridge is a graph drawn in solid steel.

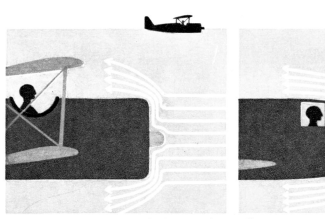

The mathematics of aircraft design takes airflow into account.

Modern aids enable the draughtsman of today to solve problems that would have baffled the wisest mathematicians of ancient times.

Small-radius compass

With the help of instruments which look simple, a young draughtsman or engineer's apprentice can now solve problems that would have baffled the most learned mathematicians of antiquity. With a slide-rule, much improved since Oughtred invented it in 1621, he can find the area of any circle and the square or square-root of any number with sufficient accuracy for his purpose in a few seconds. With a micrometer, he can measure the thickness of a piece of metal to within one ten-thousandth part of an inch. With a protractor he can lay out any angle with even greater accuracy than the priests of Egypt could lay out a right-angle. With the help of French curves, he can trace out graph outlines beyond the scope of the ruler-and-compass geometry of Euclid.

Steam power and electric power have freed our muscles from a great deal of hard, tiring work. New mathematical tools have also freed our minds from the drudgery of much time-consuming calculation. To re-draw a ground-plan on a scale three times larger than the original, a draughtsman of bygone times would have had to measure each line carefully and to multiply its length by three before re-drawing it. The draughtsman of today simply adjusts his proportional dividers so that the distance between one pair of points is three times as great as the distance between the other pair. When he sets one pair of points to the length of a line on his original drawing, the other pair automatically then shows what length the same line should be on the new one.

In the age of Newton, mathematicians had already equipped the astronomer and the engineer with log-tables which enabled them to turn problems of multiplication or division into the much simpler operations of addition or subtraction. In the age of power we have electronic calculators which can solve the most complicated problems of arithmetic in the twinkling of an eye.

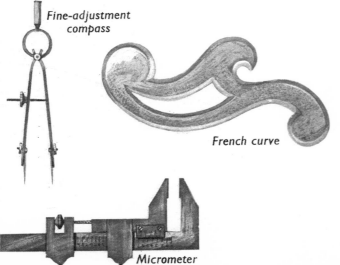

Fine-adjustment compass

French curve

Micrometer

Proportional dividers embody in brass and steel the whole idea of ratio.

Of course, it would be foolish to imagine that we are wiser than our forefathers merely because we can calculate faster than they could. The very aids which enable us to do so are based on knowledge which past generations have discovered for us. If no one had ever worked out an accurate value for π, we should not now have slide-rules capable of helping us to find the area of a circle. If no one had ever learned to divide the circle into degrees, we should not now have protractors to help us to lay out angles.

Even when we use the electronic calculator we are indebted to the long-forgotten eastern merchant who first adapted number signs to the layout of the abacus. His predecessor, the temple scribe who gave to each pebble a number value ten times as great when moved one groove to the left, first gave ordinary men a clear idea of the use of a fixed base in mathematics. The electronic calculator of today still makes use of a fixed base, though it commonly employs a base of two instead of ten. With a base of ten our columns from right to left stand for ones, tens, hundreds, thousands and so on. If the base is two, they stand for ones, twos, fours, eights and so on. When we use a base of two, we can write any number with the help of only two signs, one standing for one and the other for zero. In our diagram below we use + for one and − for zero, but other signs would serve equally well.

All our modern aids to calculation are the rewards of work done in the past. But the mathematicians of the age of power are using the heritage of the past to forge new tools of scientific thought for the use of future generations.

By challenging one of the few points which Euclid took for granted, and by convincing himself that it need not be taken for granted, Karl Gauss, a great mathematician of last century, founded an entirely new system of geometry which helps the astronomer to calculate the distance of remote stars. With the help of a calculus, different from that used by Newton and Leibniz, Albert Einstein, the greatest mathematician of our own century, worked out his famous theory of relativity which helps the scientist to a better understanding of the inside of the atom and the movements of the stars. If we look at one of Einstein's equations, $M_v = \dfrac{M_0}{\sqrt{1 - \dfrac{V^2}{c^2}}}$, we can see how his great work depends on the numerals and working-signs of earlier ages.

And so, step by step, progress in mathematics continues. It may well be that the future holds in store even greater discoveries than any yet made since the far-off days of the first moon-calendar.

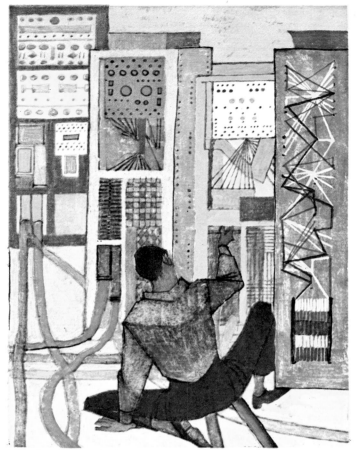

The electronic calculator often uses the simplest base of all: 2.

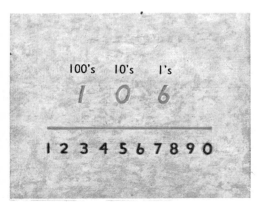
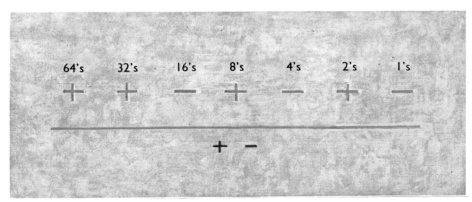

When we use a base of 2, we can write any number with two signs. In the diagram above, + stands for 1, − for 0. The number is 106.

The Beginning

25,000 B.C.
to
5000 B.C.

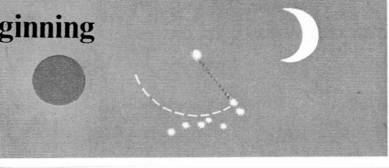 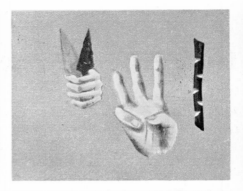

Direction-finding from stars about 23000 B.C.
Tilling the soil begins around 6000 to 5000 B.C.

Ancient Egypt

5000 B.C.
to
500 B.C.

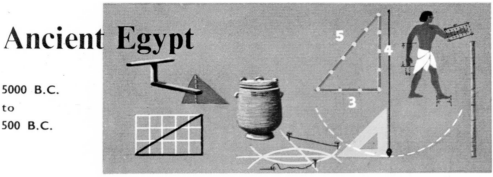 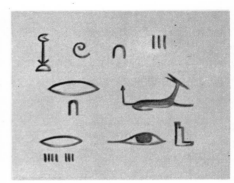

Sun-calendar possibly 4241 B.C.   Great Pyramid about 2900 B.C.
Papyrus of Ahmes about 1600 B.C.   First sun-dials about 1500 B.C.

Babylon and Assyria

5000 B.C.
to
500 B.C.

 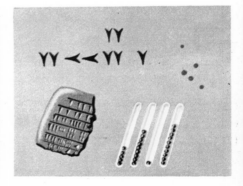

Wheels used by 5000 B.C.   Tables of eclipses by 2700 B.C.   Clay tablets with measures
about 2400 B.C.   Tables of squares about 2200 B.C.   Stamped silver bars by 650 B.C.

Phoenician Voyages

1600 B.C.
to
500 B.C.

Carthage founded 813 B.C.   Voyage around African
coast 700 B.C.   Alphabetic inscriptions by 600 B.C.

Greece and Rome

800 B.C. to 450 A.D.

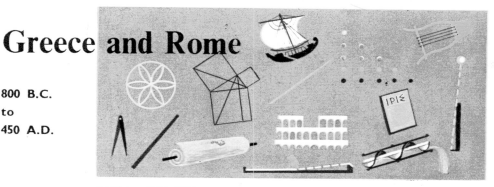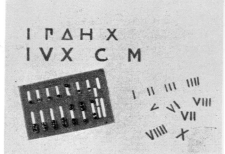

Thales in Egypt by 600 B.C.　　Pythagoras's Brotherhood 530 B.C.　　Alexandria founded 332 B.C.
Euclid's Elements 300 B.C.　　Archimedes 287–212 B.C.　　Roman Empire in its prime by 50 B.C.

Moslem Empire

640 A.D. to 1250 A.D.

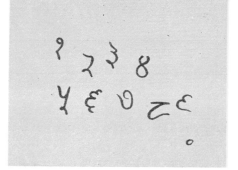

Moslems conquer Persia and Egypt by 640 A.D., Spain about 720.　　Baghdad founded about 760.
Hindu numerals in use by 766.　　Universities in Spain by 800.　　Adelard's translations 1120.

Western Europe

1250 A.D. to 1775 A.D.

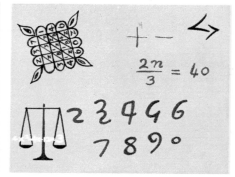

New numerals widely used by 1400.　　First printed arithmetics by 1480.　　Columbus in New World 1492.　　Galileo 1564–1642.　　Descartes 1596–1650.　　Newton 1642–1727.　　Leibniz 1646–1716.

The Industrial World

1775 A.D to Today

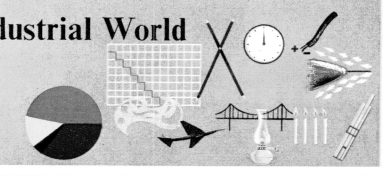

$$M_v = \frac{M_0}{\sqrt{1 - \frac{v^2}{c^2}}}$$

Watt's engine in use by 1780.　　Gauss 1777–1855.　　Gas street lights 1805.　　Steam trains 1825.
Electric lights 1876.　　Automobiles 1885.　　Wright brothers' flight 1903.　　Einstein 1879–1955.

The Wonderful World of Communication

CONTENTS

  6   **Art to Alphabet**

20   **Penman to Printer**

40   **Magic Lantern to Movie**

48   **Telegraph to Television**

56   **Our One World**

Designer: Pamela Holmes
House Editor: Josephine Waldron

Produced by Rathbone Books, London · Printed in Great Britain by L. T. A. Robinson, Ltd., Lo

Lancelot Hogben

the wonderful world of communication

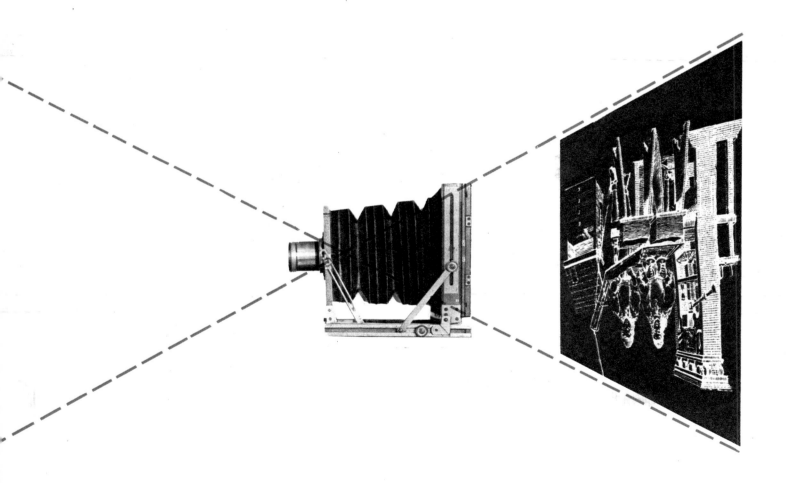

© 1959, Lancelot Hogben and Rathbone Books Limited
Library of Congress Card Catalog No. 59-10234

GARDEN CITY BOOKS   GARDEN CITY   NEW YORK

Because modern man is more talkative than the near-men who lived before him, he is better able to hand on experience to the young.
N. R. Farbman – LIFE (c) 1955 Time Inc.

Art to Alphabet

Maybe more than 200,000 years ago, there first lived on our earth beings not as yet human, but far more like ourselves than any great ape now alive in the tropics. They had bigger brains than such apes have. They could chip stones to make crude tools or missiles for hunting. While they chipped, sparks set light to dry rubble and in time accustomed them to a glow which they could fan into warm flames in the cool of the evening. In short, they learned how to make fire and not to run away from it, as do other creatures of the wild.

To say that they could make fire is also to say that they had found out, unknowingly, how their descendants could carry an artificial climate with them as they spread into the colder parts of the planet on which we ourselves now live. To say that they could make tools of a sort is to say that they had made the first step towards learning how to change their surroundings in a way which other creatures have not the power to do.

They have left behind them no remains but their bones, their chippings, charred relics of their fires and, in one sense, ourselves. Among their descendants who died out some 20,000 years ago were creatures far more like us; but they walked with a more shambling gait than ours, and their jaws tilted backward. Such near-humans made far better tools than their less human ancestors. They had stone axes and wooden spears for the hunt, rough dwellings to protect them, and hearths beneath which they buried their dead. Probably they could give simple orders and warnings by the cries they made; but what we know about the shape of their jaws leads us to doubt whether they could make as many sounds as we can.

People like ourselves turned up about 25,000 years ago. There are many local varieties of them, but they all have this in common. From the beginning, they were better craftsmen than the near-men, and they walked more uprightly. They were more talkative, and thus better able to teach their young how to handle tools.

From the start, we know more about them than we know about their predecessors because they could make pictures in colour on the walls of caves where they congregated for magical rites. In painting such pictures they had unknowingly taken the first step towards the power to communicate beyond reach of the human voice and beyond the grave.

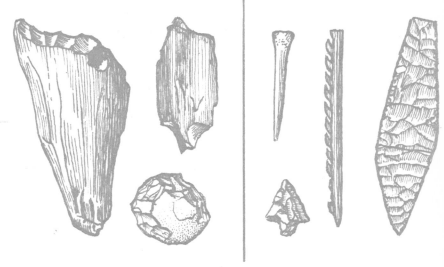

Near-men made only crude tools. Modern man, learning from his elders, made better, thus extending control over his surroundings.
Oakley, Man The Toolmaker, British Museum (Natural History)

Bataille, Lascaux, Editions d'Art Albert Skira
In making these cave paintings, man took a first step towards being able to communicate beyond reach of his voice and beyond the grave.

Dawn sky in March,

June,

September,

December. (All at 52° N.)

Pierpoint Morgan Library

Early man named star-clusters after tribal animals, which also provided his first clothes. He notched sticks to record the passing of days between the dawn-rising of one cluster and the next. The zodiac bears witness to his work as a calendar-maker.
Radio Times Hulton Picture Library

In the remoter parts of Africa there are still people who make rock drawings much like those of the first people of our species. So we have a clue to why the earliest artists, working in the depths of remote caves by the flare of crude torches, painted animals. Almost certainly each animal they painted was an animal sacred to a tribe or clan, an animal never hunted for food save at the annual sacrifice, dated by the dawn-setting or sunset-rising of some bright star or star-cluster.

Assuredly, the people of the cave paintings lived in small clans, moving from place to place in search of whatever food was abundant in successive seasons. Certainly also they knew far more than most city dwellers of today know about the rising and setting positions of the stars on the horizon, the changes of the night sky, the phases of the moon and the changing track of the sun across the heavens in the course of the seasons.

Having no calendars, their only way of keeping track of the seasons was to notice and to remember which star-clusters rose just before dawn or set just before sundown at different times of the year. Having no maps, their only way of charting their trek to where fruit, seed grasses and game would be abundant was by the star landmarks with which they became familiar during long fireside watches under the night sky.

Very early they must have learned to count the days between one new moon or one full moon and another. By doing so, they were better able to plan ahead for their journeys. Gradually they also learned to realise the advantage of being able to keep track of time still further ahead. This was not easy at first. To keep track of a whole year, by counting the days between two successive occasions when the same stars set or rise just before sunset or just before sunrise is not like counting sheep in a flock. Since sheep are all in one place, we can count them a second time if we are in doubt; but we can count days only in succession. We can never retrace our steps. The book-keeping of time over the course of many years is burdensome and well-nigh impossible unless it is your main job. If it is, you soon find the need to use some artificial aid to the memory.

This is difficult if your tribe is always on the move. But where people began to drive herds into ravines with the help of dogs and to sow the grass seeds they found sprouting on return to a former camp, they had less need to wander. They had now

Long before white men first went to America the Mayas already had great temple-observatories and a remarkably accurate calendar. Each face on this Mayan calendar-stone stands for a different month. The numerals clearly indicate that they originated as simple notches.
M. von Harten

the opportunity to set up stones or poles fashioned from tree trunks and so to notice the changes of the sun's shadow throughout the day or throughout the seasons. Day by day they could chip on them marks as a reminder of how long it was since a certain star was last noticeable near the eastern or western horizon at dawn or sunset, or since the sun last rose against a particular hillock or clump of trees.

By doing so, they took the first decisive step towards writing. We can guess with good reason that the second step was to use pictures of a sort to show which marks recorded which event. This is reasonable, because some early sorts of writing consist only of a number of signs which are dots or strokes (like the first three Roman numerals), and of pictures of sacred beings, each with its own calendar day of sacrifice, or each standing for a particular month of the year. We find such writing on the stone pillars of temples buried in the jungles of Guatemala and Honduras in Central America. We speak of the people who made them as Mayas.

We can be sure that man blundered into writing as men or near-men blundered into making fire or wearing clothes. How the ape-man blundered into fire-making we have seen. How our own species blundered into dressmaking, the cave paintings help us to understand. When marriage between persons of the same clan was a capital crime against tribal law (as it was among some remote tribes within the memory of people still living), a man wore the skins of the animal sacred to his own clan as a sort of passport to matrimony.

Many thousands of years passed before men realised that the use of fire and clothing would make it possible for human beings to live in any climate. Several thousand years may well have gone by before people who already had crude writing good enough for calendar-keeping (but for very little else) blundered into writing as a means of sending messages far beyond the range of the human voice and of recording experience for the instruction of pupils they were never likely to meet.

Long before writing as we know it began, the people of Mesopotamia used seals bearing animal pictures as signs of ownership. In time they developed a far wider range of signs, punched on soft clay with a wedge.
British Museum

A more settled pattern of life among herdsmen of sheep or cattle and tillers of millet or barley appeared first in Egypt and the Middle East about six thousand years ago. A priestly caste of calendar-keepers, concerned with bribing the gods by sacrifices to give favourable weather, there had leisure to study the seasons. They had also the urge to employ labour to build temple-observatories and the opportunity to exact tribute from the herdsmen and tillers. While all men lived, as some still do, in small clans wandering from place to place in search of food, clansmen had shared what little they owned; but among the priests there was now private property.

At first the tribute on which they waxed fat was chiefly grain and carcasses – goods which must be quickly consumed or else wasted. Soon, however, cities sprang up around the precincts of the temple-observatories, and merchants began to bring goods for exchange from one city to another. Money of a sort, such as precious metal stamped to guarantee its value, became a necessity of trading; and money provided the privileged priests and merchants with the means of hoarding whatever wealth they did not need for immediate use.

So we now see picture-making put to use as a means of marking ownership. Almost certainly, before there was any writing as we understand the term, it was customary to use seals to stamp its ownership, its quality or the identity of its maker on metalwork and pottery. Many of the earliest seals which have been unearthed carry pictures of animals, and it may well be that each such creature was originally, like those of the cave paintings, sacred to a particular clan. Be that as it may, the seal gives us the clue to a third ingredient in writing.

Meanwhile, the priests within their temple-observatories had a greater urge than their predecessors to record the changes of the heavens and a greater need to keep an inventory of their property. In time the priestly accountants needed far more than a few number

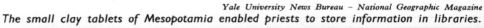
Yale University News Bureau – National Geographic Magazine
The small clay tablets of Mesopotamia enabled priests to store information in libraries.

signs and a few pictures to symbolise particular heavenly bodies. Their battery of signs, still pictorial, grew till it could convey a message recording an important historical event, an instruction for the young priest in training, or a clear working order to a vassal.

As the signs increased in number, they became less and less like the detailed pictures they once were. This was so especially in Mesopotamia, where writers adopted the trick of punching each sign with a wooden wedge on flat slabs of soft clay before leaving them to harden in the sun. When we compare such small tablets of clay found neatly stored in the four-thousand-year-old ruins of Mesopotamian temple libraries with the huge calendar stones of the far later Mayan temples, we see at once what a saving of space they represent. We then realise what an important step forward the Egyptians took when they began to write on something like paper, thin sheets of reed called *papyrus*.

By doing this, they solved for an age in which few people could read, the kind of problem still facing engineers who are designing electronic brains for accountancy and for storing scientific information. They equipped mankind with a man-made memory. For the substance on which they recorded information takes up little more space than anything men had till the coming of the magnetic tape.

British Museum
The picture-writing of ancient Egypt, on thin sheets of papyrus, made it possible to equip man with a more compact artificial memory.

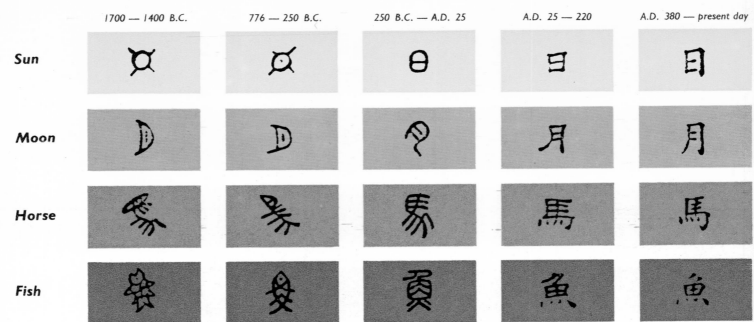

Chinese characters have gradually developed from simple pictures. They do not express the sound, but the meaning, of words.
China Reconstructs

Chinese of North and South speak differently but write alike.

	N. China	S. China	All China
🌲	Sung	Ts'ung	松
🐂	Niu	Ngou	牛
👞	Hsieh	Hai	鞋

Even when the writing of ancient Egypt and Mesopotamia ceased to have any recognisable trace of its pictorial origin, it was very different from any writing in the Western World of today. People who use the alphabet can recognise any word by the sounds which the letters represent. For written signs have nothing to do with the meaning of the word. What a set of letters in a particular order may mean to the reader depends on the speech of the writer. For example, a Frenchman writing *or* and *main* means something quite different from an Englishman who uses the same combinations of letters.

Now each word in the ancient priestly writing had its own sign; that is, each sign had a meaning which had nothing to do with sound. We ourselves use some signs of this sort. When a Frenchman says *cinq* or *et*, an Englishman *five* or *and*, and a Swede *fem* or *och*, they may be unable to understand one another; but the written signs 5 and & convey the same meaning to each of them.

The only writing the Chinese have used till now is of this sort. This is a great handicap to Chinese education. To write about difficult subjects we need many words; and if each word with a particular meaning has a sign for itself, the number of signs we need is very great. Consequently it takes a very long time to learn to read.

On the other hand, the Chinese script has had one great advantage. People in the north and in the south of China do not speak the same languages. Their languages may sound no more alike than

Sound	Parent Chinese Character	Katakana	Parent Chinese Character	Hiragana
ka	加	カ	加	か
ki	幾	キ	幾	き
ku	久	ク	久	く
ko	己	コ	己	こ
shi	之	シ	之	し
su	須	ス	す	す
se	世	セ	世	せ
so	曾	ソ	曾	そ
ta	多	タ	多	た

Bodmer, Loom of Language, Allen & Unwin Ltd.

Early Japanese painting. When the illiterate Japanese first came into contact with Chinese culture they borrowed Chinese signs to represent the sounds of their own syllables. They built up two syllabaries, the Katakana and the Hiragana. (Corresponding parts of both shown above.)
Swann, 2000 Years of Japanese Art, Thames & Hudson Ltd.

Swedish and German or French and Roumanian. Without learning how to do so, they cannot therefore understand one another when speaking. But those who know the old script can read the same classics or newspapers, because each sign, like our sign 5, has a meaning which does not depend on the particular sound we make when speaking.

There is another good reason why the Chinese have been so long unwilling to discard a kind of writing which few but scholars could ever hope to master. Nearly all essential words of Chinese languages are either simple syllables, like *bee, toe, no, me,* in English, or combinations of simple syllables, like *lay-by* or *cow-boy*. It is possible to make very few distinguishable words of so simple a sound pattern. Consequently, the same Chinese sound may have many different meanings. For Chinese languages a script which uses a different sign for each meaning is therefore less likely to cause confusion to the reader than a script which uses a different sign for each sound.

The Japanese, whose language consists of strings of simple syllables, like NA-GA-SA-KI, TO-KI-O, came into contact with Chinese culture about 300 A.D. At that date, they had no writing; and they could not yet read. When people who cannot read point to a written sign in a language which they do not understand, the writer, unable to understand their own language, can do no more than make a noise which has no meaning to them. His hearers thus come to regard the sign, like notes on a score of music, merely as a way of writing down a sound.

We have seen that Chinese words are mostly single syllables, each with several different meanings and several different signs. It is therefore not surprising that the Japanese have been able to devise from Chinese characters more than one battery of signs for all the syllables of their own language.

Photo. by George C. Cameron, University of Michigan and the American Schools of Oriental Research

The Rock of Behistun carries the same message in three languages, Old Babylonian (blue), Susian (yellow), Old Persian (red). Scholars who knew Old Persian were thus able to decipher the other two. Both use syllable-signs. Above is the name Darius in all three languages.

For telegrams and advertisements, the Japanese still use syllable-signs only; but they have now come to print their books partly in syllable-signs and partly in Chinese characters, each with its own meaning. This is because they have taken over an enormous number of Chinese words which may have several different meanings. The result is that learning to read and write is now a headache for Japanese children.

Nevertheless, if we think only of the simple kind of writing which the Japanese use for telegrams, we have the clue to one way in which writing has ceased to be the possession of a privileged few, and can now be the birthright of every citizen. If the speech of a people has only 15 consonant and 5 vowel sounds (as is true of Japanese), we can make only 15×5 syllables of the type TO or KI and 5 of the type O in TO-KI-O. With 80 signs we can thus represent all the sounds we use, provided all our words are of this pattern.

The words of many languages are, in fact, of the same simple pattern as TO-KI-O and YO-KO-HA-MA, and where the range of consonants and vowels is small, the number of possible syllables is also very small. In one of the languages of the Philippine Islands there are only 52 syllables, and in Hawaiian only 40. As soon as you have learned to recognise the signs for these 52 or 40, you have completely mastered the art of reading.

So it need not amaze us that many peoples of the world and at different times – in the Middle East, in Crete, in Cyprus, in the islands of East Asia, and in West Africa – have mastered the trick of writing by syllable-signs.

More than a century ago, near Behistun in Persia, archaeologists found an inscription carved in three ancient languages. From what was already known of one language, Old Persian, they were able to decipher the other two, Susian and Old Babylonian. Both were written in a script made up of signs for simple syllables.

Far more recently, a young scholar named Michael Ventris spent a great part of his life patiently decoding one of the long-forgotten scripts of ancient Crete called Linear B. Like many other scripts in various parts of the world, this also consists of a battery of something over seventy syllable-signs.

In short, wherever people speak languages with a sufficiently simple sound pattern it is not hard to stumble on a system of syllable-writing. But a little arithmetic will convince you that writing by syllable-signs would be of no use to people who speak languages such as Greek or English which make use of syllables such as *strands* or *trips*. With 24 English consonant sounds and at least 12 vowel sounds, we could build up $24 \times 24 \times 24 \times 12 \times 24 \times 24 \times 24$ syllables of the first type and $24 \times 24 \times 12 \times 24 \times 24$ of the second. To be sure, a sizeable proportion would be unpronounceable. Even so, the number of syllables employed in English, Russian,

These Cretan writings are more than 3,000 years old. The script employed, called Linear B, defied decipherment until recent times.
Ashmolean Museum

Greek or French words runs into thousands, each of which would require a separate sign if we used syllable-writing.

To spell such words with a small number of signs is possible only if we break down the syllable itself into its smallest bricks – the consonants and vowels. Writing by alphabet-signs for the separate consonant and vowel sounds in a syllable has turned up once and once only in history. The possibility is not by any means obvious; and it is doubtful if it could have happened in more than one way.

Camera Press, Ltd.

Michael Ventris spent some years patiently decoding Linear B.

L. Ventris
It proves to consist of a battery of seventy-odd syllable-signs.

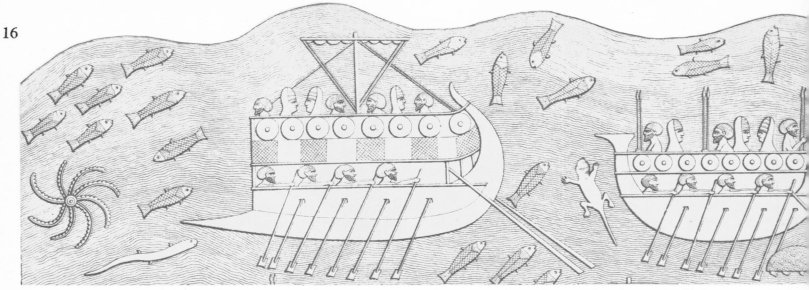

The Semitic Phoenicians were great sea-traders. Before 600 B.C. they spread the use of letters far afield in the Mediterranean World.

The Metropolitan Museum of Art

In brief alphabetic inscriptions, some early scribes wrote left to right, some right to left, others both ways or even in a spiral. Some carved on rock, others painted on pottery. Letters varied in shape according to the writing tool and the hardness of the surface.

British Museum

The story of the alphabet begins with a Semitic people seemingly enslaved about 1800 B.C. by alien conquerors in the mines of Egypt, like the people of Israel in the Bible story. They spoke a language akin to the Hebrew of Israel and to the Arabic of Jordan today. Such languages have a sound pattern of their own.

Aside from terminals such as *un-* in our word *un*kind or *-ed* in our word polish*ed*, most Semitic words consist of three consonants with two vowels sandwiched between them, as in DAVID. The vowels may change, as in our words m*a*n and m*e*n, f*oo*t and f*ee*t, g*i*ve and g*a*ve, without affecting the essential meaning. So you know the essential meaning of words such as our *salad, demon, tepid,* if you write them as s.l.d., d.m.n., t.p.d. Thus the first alphabets, used for writing similar Semitic words, needed only consonant signs. These were probably copied from signs for Egyptian one-syllable words starting with the same consonant. Even now, Semitic alphabets have only dummy vowel sounds.

The Semitic Phœnicians, who were the great seafaring traders of antiquity, spread the use of letters far afield in the Mediterranean world before 600 B.C. There, Greek-speaking peoples, who had already tried to use syllable-writing in Crete and in Cyprus, by then had vowel signs to add to the B C D alphabet of their teachers. In the Greek language vowels really matter, just as they do in such English words as m*a*n, m*oo*n, m*ea*n, m*oa*n, m*i*ne and m*a*ne.

Though they all have one Semitic parent, many alphabets now in use have lost all trace of family likeness. They began to lose it at an early date. When people first used alphabet-writing, they did so only for small jottings and short inscriptions. So it was not very important to write in a straight line or in any fixed direction. When they did settle down to a straight line, some writers wrote horizontally and some vertically, some from left to right as we do, others from right to left, as in the earliest Oscan of Italy, still others in alternate directions. In the course of these changes, some letters settled down sideways, left or right, others on their backs.

Meanwhile, people of different countries used different writing materials such as papyrus, wax, stone or wood and different writing tools, such as brush, quill, stylus or chisel. It is not as easy to make a smooth curve on stone as it is on wax; not as easy to make a sharp angle with a quill as it is with a chisel. So, in different regions in the course of many years, letters took different shapes dictated by what people used to write on or to write with.

Tyrian, c. 900 B.C.	Early Greek, c. 750 B.C.	Later Greek, c. 400 B.C.	Roman, c. 300 B.C.	Russian Today
△	▷	△	D	Д
∧	∧	Γ	C	Г
⇁	K	K		К
ᓂ	∨	∧	L	Л
ᗰ	ᙢ	M	M	М
Y	ᐱ	N	N	Н
⟩	Γ	П	P	П
⊲	Ʀ	▷	R	Р
✕	T	T	T	Т

The alphabet was invented once only. Over many years different forms evolved in different areas. But we can still see a strong family likeness, especially if we look only at D, G, K, L, M, N, P, R, T.

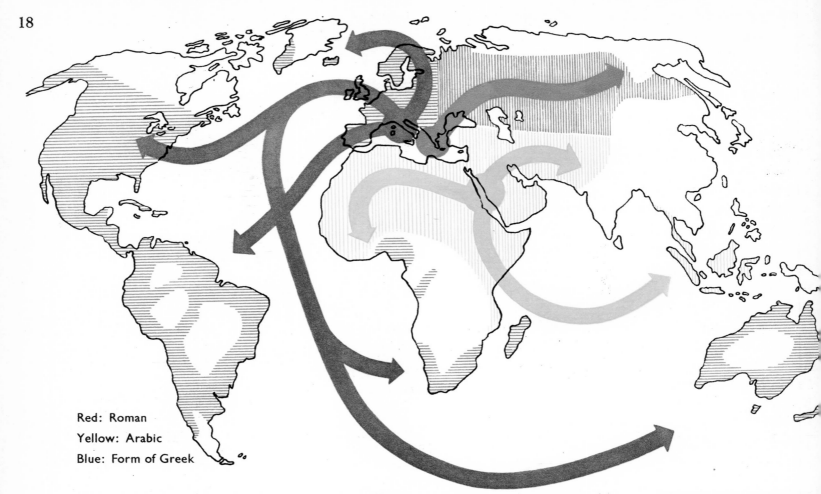

Red: Roman
Yellow: Arabic
Blue: Form of Greek

Missionaries of the Western Church spread the Roman alphabet, those of the Greek Church a form of the Greek, those of Islam the Arabic.

During close on four thousand years since alphabet-writing began, traders and missionaries – Buddhist, Roman Catholic, Greek Orthodox, Moslem, Protestant – have made one form or other of the alphabet available to people using vastly different languages.

People who do use different languages do not necessarily use all the same sounds. For instance, the French do not use either of the sounds which English-speaking people represent by *th;* the Icelanders use both, and have two different signs for them, þ for the sound in *th*in and ð for the sound in *th*en. Thus an alphabet tailored to the needs of one language may provide no signs for certain sounds in another. It may also provide signs for which another language has no need.

There are several ways of sidestepping this difficulty. One is to give unwanted signs a new sound value, like the *x* used for the click sound in the spelling of the South African *Xosa* language. A second is to combine old signs to meet new needs, as with *th* in English spelling. A third is to borrow signs from another alphabet, such as the Greek K, Z and Y added to our mainly Roman signs. Yet another is to make up new signs, like our W and J. All such changes make the alphabets of the world look less and less alike.

Furthermore, languages change in different ways in different regions. For instance, in some regions the V-sound now occurs in all words where the B-sound once occurred. Since spelling changes more slowly than speech, what was once the sign of a B-sound may thus come to stand for the V-sound, as in the Russian alphabet.

The study of sound changes makes it possible to notice between words in different languages many similarities that might otherwise pass unnoticed. By such means scholars are able to classify languages as members of families with the same parent in the sense that Latin is the parent of French, Italian, Spanish, Catalan, Portuguese and Roumanian.

Changes such as we have seen had been going on for a thousand years before the alphabet got into the hands of Greek-speaking peoples. Throughout that time people had used it only for short jottings. The seafaring Greeks, who learned from the Egyptians to use papyrus, now put it to much wider use.

Greek master pilots, who navigated by the stars, were immensely interested in astronomy. When they became aware of the wealth of information stored in the temple libraries of older civilisations, they were anxious to re-write it in their own tongue. The free Greek citizens, who had developed the tribal dance of their forebears into the beginnings of drama, began to write down the words of the performers.

There were thus two new uses for alphabet-writing and for writing at length. One was to accumulate scientific knowledge. The other was to improve the theatre. In fact the dialogue of the drama became the model for writing down the record of philosophic disputes and the words of renowned teachers. To an extent never before in history, human beings began to write in a conversational way. Far more important than what they thought about science, law and philosophy is the fact that the Greek-speaking peoples thus made alphabet-writing a powerful instrument for recording what people think on any topic.

Nine	English
Neun	German
Neuf	French
Nueve	Spanish
Nove	Italian
Naw	Welsh
Negen	Dutch
Nio	Swedish

Similarities between words enable linguists to group languages into families. A likeness between Indo-European languages is often clear in words beginning with N. Other likenesses appear when we remember that many sounds shift consistently, as may S to H or P to F.

Greek	Latin	Greek	Swedish
Hexa- *(six)*	**S**exa-	**P**olos *(foal)*	**F**åle
Hemi- *(half)*	**S**emi-	**P**odo- *(foot)*	**F**ot
Hepta- *(seven)*	**S**eptem	**P**ente *(five)*	**F**em

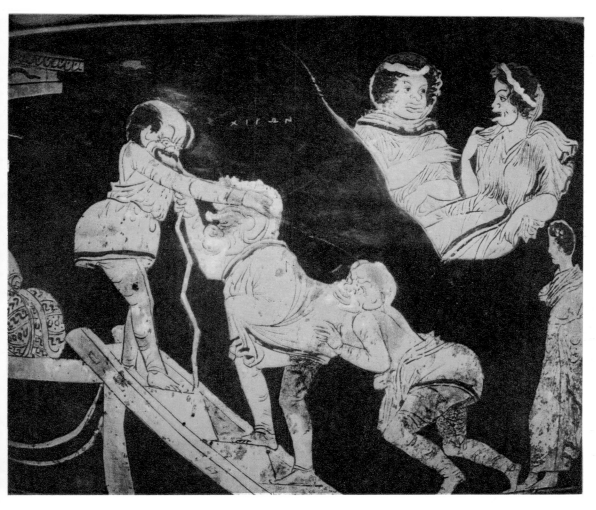

This vase, showing a scene from Greek drama, reminds us that the Greeks were the first to use the alphabet for writing at length, in plays and in scientific works.
British Museum

British Museum
Until the end of the Middle Ages few people could afford to read. Duplicating a book demanded the slow work of the scribe. Writing material was so costly that people often used the same piece several times.

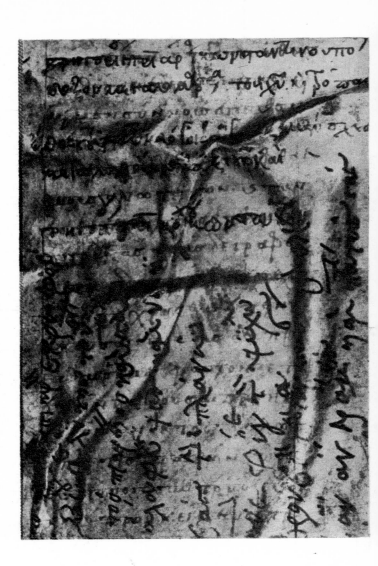

Penman to Printer

For two thousand years after Greek-speaking people took over alphabet-writing, reading was still an art few people could enjoy. Still fewer could write. Books, which could be copied only by hand, remained an expensive luxury.

After Roman armies destroyed Carthage and later annexed the Greek mainland, Greek-speaking peoples, especially in the great sea port of Alexandria, continued to dispute, to discover and to write. Meanwhile Roman militarism dominated Old-World civilisation west of Persia for six centuries and added nothing to scientific knowledge.

As a centre of learning Alexandria passes out of history about 400 A.D. Some three hundred years later, Semitic followers of Mohammed began to overrun Asia Minor, Persia, North Africa and Spain. Within this empire they harvested not only the achievements of the Greeks but also the new number script and the new arithmetic of the Hindus.

In Spain they founded universities at Toledo, Cordova and Seville, where mathematics, astronomy, geography and medicine flourished. From these centres, Jewish physicians, traders, and monks who came in disguise gradually spread the knowledge of Western and Eastern antiquity throughout Europe during the twelfth and thirteenth centuries A.D.

Before that, little of the learning of the ancient world survived in Europe, and that little only in Constantinople and in sunny Italy, where wealthy merchants traded with the East. In 1200 A.D. wellnigh the only people in North-West Europe who could read were priests, lawyers, a few master pilots and Jewish traders. The wealthiest lived in stone castles with glassless open slit windows.

Reading during the dark winter days was possible only by candle or rush light.

Two centuries later, the picture had changed. England and Germany were prospering by trade. The homes of the wealthier merchants and craftsmen had windows with leaded glass. There were grammar schools founded to give instruction in Latin, still the language of Church and Law. In some, sons of prosperous townsmen were beginning to learn the new Hindu-Arab arithmetic.

First in Italy, then in Britain, France and Germany, chroniclers and poets, such as Dante, Petrarch and Chaucer, began to write in the native tongue. Wycliffe made the first complete translation of the Bible into English. A few manuals of navigation circulated among master pilots, and manuals of the new arithmetic in the counting-houses.

The fourteenth century saw schools opening for merchants' sons.
Biblioteca Angelica – Salmi, Italian Miniatures, Electa Editrice
British Museum

Throughout Western Europe, poets and chroniclers began writing in their native tongues.   Wycliffe translated the whole Bible into English.

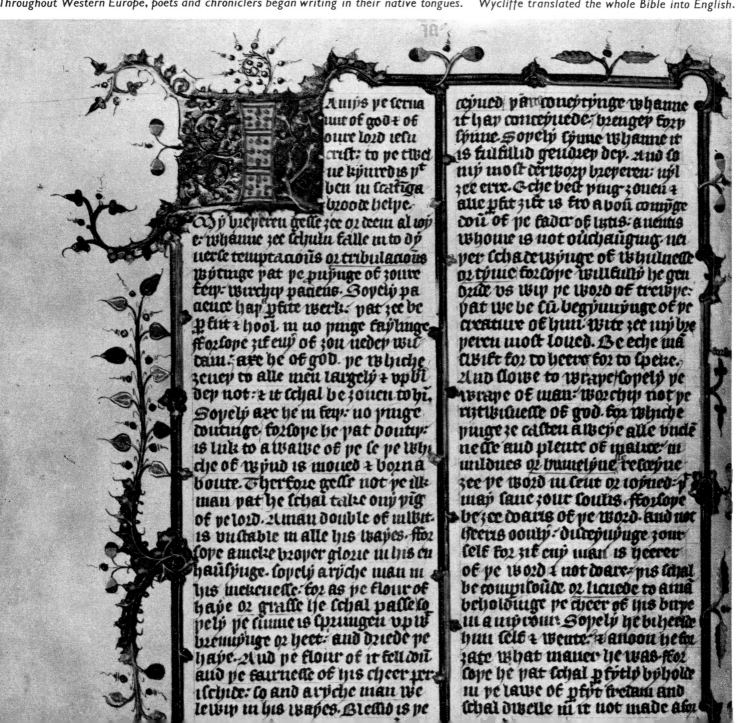

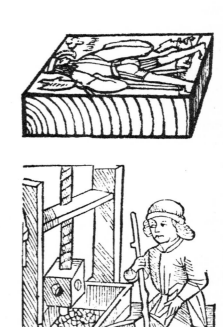

Hofmuseum, Vienna

At the time printing from movable type began, many of the tools were already to hand. The wine trade had long used presses not unlike printing presses. Paper-making had gone on in Europe for two hundred years. The wood-block printing of playing cards was already in full swing.

John Omwake Collection, Cincinnati Art Museum

Meanwhile, copies of any book were scarce. Few people could afford to possess them. Available reading matter was the workmanship of monks or lay scriveners, copied word by word with a pen. About fifty years before printing from movable type began, an Italian prince hired forty-five copyists to make his private library. By working hard for nearly two years, these men produced only two hundred volumes. Production of a single copy of one book would keep one scrivener employed for a period of several months.

It was also costly. Before 1200 A.D. monks and scriveners in European countries had to use parchment prepared from animal membrane; but the Chinese had long since discovered how to make a better and cheaper writing material. This they did by compressing textile waste, such as rags and old rope, after seeping and softening in water.

From China, paper-making spread through India into Persia and throughout the Moslem Empire.

Scholars in contact with Moslem universities in Spain and traders who exchanged goods with the Near East spread throughout Europe the news of this invention, and of another. Early in our era, the Chinese began to exploit a very old device in a new way. They used wooden seals first to stamp designs on textiles, then to stamp pictures on cards used for games of chance. Later, they and their Japanese pupils used wood blocks to depict the great Buddhist teachers. From this it was but a short step to stamping writing-signs.

By the time of Petrarch and Wycliffe, wood-block printing of playing cards and of pictures of saints was current in Europe. During the following century, a few books in circulation had text printed off from wood blocks, each block carrying the text for a whole page. The men who made such blocks had to carve separately every letter of every page.

Skilled goldsmiths and armourers already held one clue to a much speedier method of printing. They understood how to make from a single punch or die many copies of individual letter-signs in metal. These the printer can temporarily assemble to make a page-size block. But the recipe for printing from movable metal type was incomplete until the discovery of a suitable ink.

The great painters of the fourteenth century had used water paints, stiffened as required with egg-white, and water pigments had been the only ink of the wood-block printer. Though such inks cling to wood, they run off metal.

During the fifteenth century, Flemish and Italian painters had learned to work with oil paints. Those in touch with metal craftsmen could produce many copies of an engraving by cutting, on a flat metal plate, crevices which could hold an oily ink.

With thick, oily ink of the kind these artists used, it was now possible to print from metal type, and thus to produce hundreds of copies of a book while one scrivener could complete little more than one.

Some forty years before Columbus set out on his most fateful voyage, a German master-printer of Strasbourg named Gutenberg succeeded in printing from movable type the first book of its kind. In 1467 German printers brought the art to Italy. A year later there was a printing press in Paris. In 1475, Caxton, then working in Bruges, had his own translation of a medieval legend printed there. Next year he returned to London, where he set up an English printing house near Westminster Abbey.

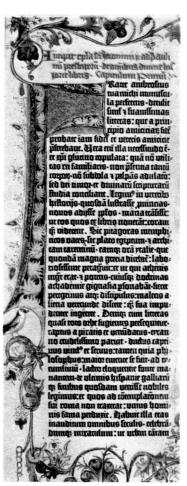

Skilled goldsmiths and armourers became the first type-founders. This page from an early printed book bears witness to the high quality of their craftsmanship.
British Museum – Mansell Collection

Bibliothèque de L'Arsenal
The products of the early presses met with a ready market. Merchants were anxious for books that would explain how to make use of the new Hindu-Arab arithmetic in their counting-houses.
British Museum

Seamen, now venturing into the open Atlantic, also needed books as never before. They wanted forecasts of eclipses, for finding longitude, and information on strength and direction of winds.
The Trustees of the National Maritime Museum

The printing presses which went into action during the forty years before the fleet of Columbus took sail for the New World had many ready buyers for the wares they produced.

Merchants who had now some schooling were eager for books that would help them to take advantage of the new Hindu-Arabic arithmetic in their counting-houses.

Sea captains, too, needed books as never before. Hitherto, knowledge of latitude had sufficed for European long-distance navigation northward to Iceland or southward along the coast of Africa. Westerly navigation into the Atlantic thus raised the problem of charting a course by longitude as well as by latitude. Seamen thus needed nautical almanacs to forecast the time of eclipses and other celestial events on which they relied to locate their east-west position.

Soldiers as well as sailors were among the printer's customers. The Western World had lately taken over the hitherto harmless Chinese invention of gunpowder; and military commanders needed textbooks of instruction in a new art of artillery marksmanship.

In Christian Europe there were now universities in which Jewish teachers familiar with the learning of the Moslem world had founded medical schools. There were also hospitals where monks and nuns ministered to the sick. Such people cultivated herb gardens, and needed books with illustrations to make clear the identity of herbs with supposedly healing properties.

From 1378 to 1417 there had been rival popes in the Catholic Church, and long before printing began the princes and prosperous townsmen of England and Bohemia were willing to listen to Church reformers such as Wycliffe and John Huss. By the time the first printed Bibles appeared there were many devout laymen eager to search the Christian scriptures as the highest court of appeal in matters of faith.

Western technology, no longer held in check by abundance of slave labour, was advancing on a wide front. Where there was deep-shaft mining, the operatives were free men who demanded a solution to the problems of flooding and suffocation underground. Among the master craftsmen of the towns, there were clock-makers facing new mechanical problems and a growing number of spectacle makers tackling new problems of optics.

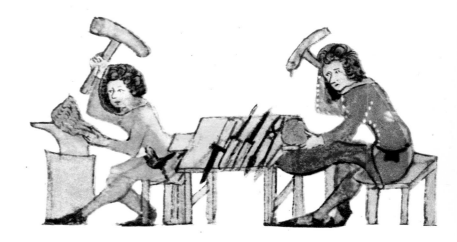

Metal workers were eager for knowledge about the recent advances in technology. Agricola's famous De Re Metallica used realistic diagrams to promote the use of improved techniques.
Bodleian Library, Oxford

A—Wood.   B—Bricks.   C—Pans.   D—Furnace.   E—Crucible.   F—Pipe.   G—Dipping-pot.

For all such men, the only high road to skill had hitherto been through apprenticeship. The only link with the accumulated experience of the past had been a tradition handed on by example and word of mouth. Printing now made it possible to spread the use of human skills to those who had no traditional links with the past.

From the beginning, printing was important because it made the written word available to thousands who could otherwise never have benefited from it. In a world bristling with new problems and new ideas, it was even more important because it enlisted the skill of the artist – a skill as old as the cave paintings – to pass on knowledge.

A century before the time of Columbus, knowledge of geography in the monasteries of Christendom was vastly less than it had been in the schools of Alexandria a thousand years earlier. Monastic maps may have had some religious significance but they had little or no relation to the real world.

But Henry the Navigator, Prince of Portugal, was training sea captains in the Moslem art of map-making while Gutenberg's press was still in the experimental stage. When these men and their successors returned from their great voyages of discovery they brought back reliable information which could be transformed with the help of the artist into printed maps which were of real use to other seafarers.

Before the coming of printed books, all that anyone could learn about the mechanical devices of earlier civilisations was from wordy and pictureless descriptions written by gentlemen-scholars who had never made or used them. All that medical students, who as yet had no access to laboratories or dissecting rooms, could learn about the structure of the human body or of the identity of supposedly healing herbs, they must need learn from hand-copied manuals with neither diagrams nor realistic pictures. Art had scarcely begun to pay dividends.

This typical medieval map bears very little relation to reality. Jerusalem looms large in the middle. Strange figures people unknown lands.
British Museum – Mansell Collection

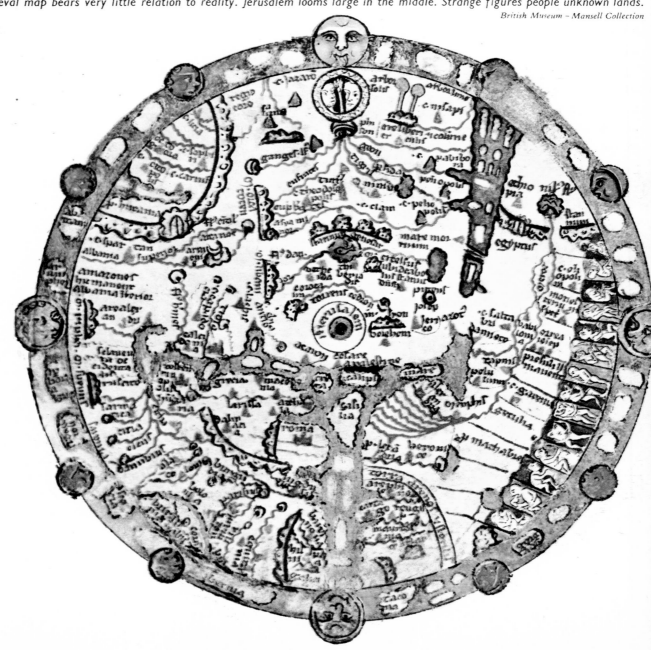

Without printing and without the artists who had done so much to help make printing possible, the great advances of medical knowledge in the following century could never have happened. Hitherto teachers of anatomy and botany had interpreted the written word as best they understood, or misunderstood, it. Now students could learn from treatises illustrated with pictures which spoke for themselves.

The great artists who took a hand in making pictures of the details of human anatomy and of the structure of plants for printed reproduction thus helped science to develop a more critical attitude. Teachers and students began to discover errors and uncertainties in the written word of the ancient authorities on whom they had so long relied. They turned more and more to dissection and to the use of their own eyes to discover what animals and plants are in fact like.

They recaptured the curiosity of the early Greeks. At a time when master printers were foremost among men of learning, it was an Italian master printer who trained the scholars who spread the knowledge of the Greek language through Western Europe; and the study of Greek literature added to the ferment of men's minds. It encouraged religious leaders to re-examine the teaching of the early Christian Church. It also prompted political reformers to look at the social institutions of their time against the background of earlier controversies. It also encouraged a growing number of authors to write in their own languages and in an easy conversational way without recourse to pretentious expressions.

The master printers of the sixteenth century also contributed not a little to the great musical revival which began with the introduction of stringed keyboard instruments. Organists of the monasteries had already committed their chants to manuscript in a crude staff notation. Printing made it possible to distribute many copies of a composer's work to those who could play an instrument.

Salmi, Italian Miniatures, Electa Editrice

A medieval drawing might represent any one of a dozen plants. An illustration from Gerard's Herbal is unmistakable.

Bodleian Library, Oxford

Early printed maps were based on information brought back from the great voyages of discovery by sea-captains skilled in map-making.

British Museum

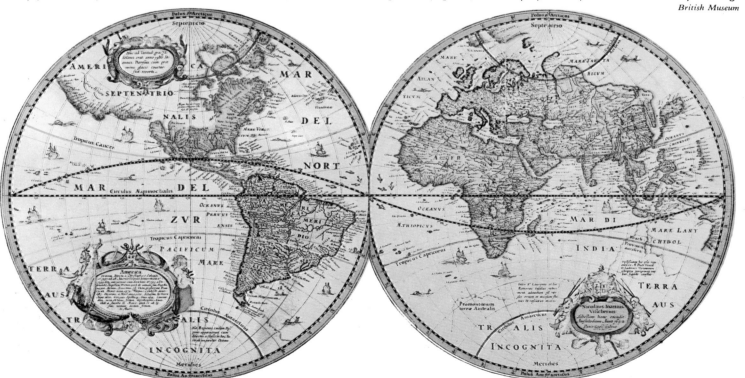

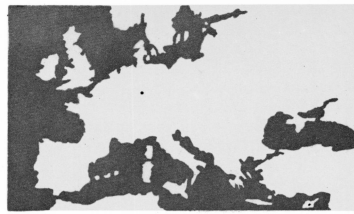

In 1450 Johann Gutenberg set up Europe's first printing press.

Some seventy years after printing from movable type began, a partly religious, partly political, tract written in German by Martin Luther sold no less than 4,000 copies in five days. A century earlier, it would have taken a thousand men to produce so many copies in the same time.

The variety and output of the printing presses during the first hundred years of their existence is not hard to understand when we recall the circumstances of sixteenth-century Europe. Lately troubled by disputes within the Church and torn by wars in which gunpowder had revolutionised strategy, it was seething with political and religious controversy. Beginning to adjust itself to the use of wind or water power to perform tasks hitherto carried out by slave labour, it was also enriching itself immensely by exploring and colonising territories on other continents. It was thus natural that printing spread rapidly throughout all Europe and its colonies in the New World.

But we may well ask why it exerted far less effect on the Far East, where printers actually did learn to make movable type of a sort from wood blocks. A true answer to this can hardly be simple. One reason is that the political régime was more stable, because the East had no insurgent class of newly prosperous craftsmen and merchants. Another is that there was no challenge of navigation in uncharted oceans. No less important is the fact that China's clumsy picture-writing still made it impossible for more than a privileged few to become proficient in reading and writing.

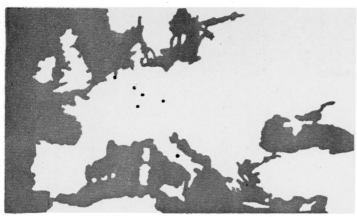

By 1465 printing had begun in Holland, in Italy and in France.

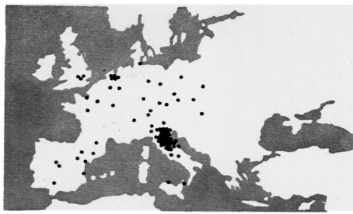

Fifteen years later there were presses all over Western Europe.

By 1500 their number had more than doubled. By 1638 Spanish, Portuguese, Dutch, French and English overseas territories had presses too.

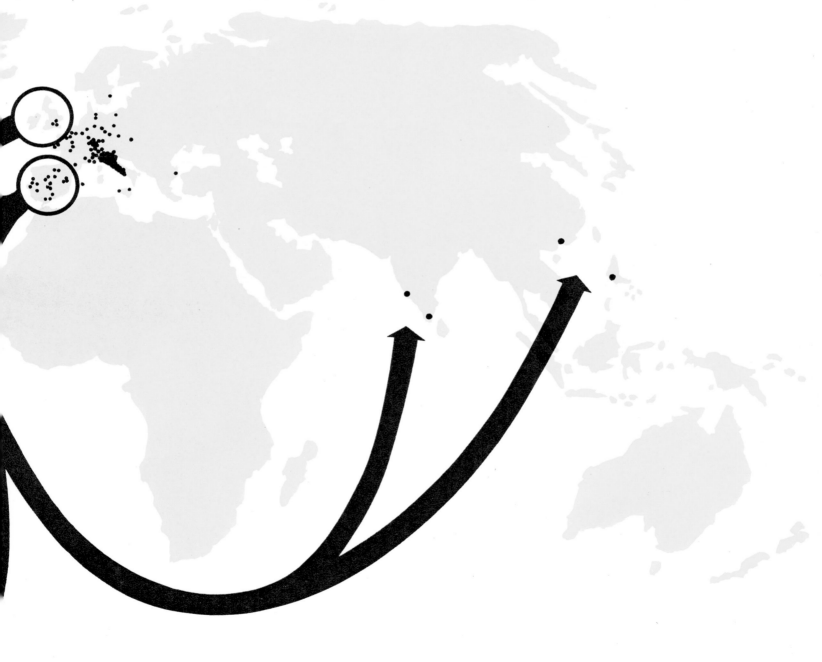

In modern industrial towns we regard ability to read the pay check or to write down the price list as essential. Indeed, we look on those who cannot learn to do so as misfits in need of protection.

This is a quite recent viewpoint. Though printing made it speedily possible for many more people to read many more books, the great majority, even in prosperous Western countries, were still letterless at the time of the French Revolution. Among peasants and craftsmen, few could read, and the simpleton could employ himself or herself usefully about the house or about the farm with little or no inconvenience to anyone.

Even educated people read slowly by our standards. When a new book was still a treasure, it was common for one of the family to read it aloud to the others by lamp-light or candle-light during the long winter evenings.

Sunday schools, started in England and Wales during the eighteenth century, gave farm workers what little instruction in reading they then had.

Not until rural craftsmen migrated in great numbers to work in the steam-driven factories of the growing towns did Western countries begin to have free schools or schooling that all could afford. Steam power resulted not only in a greater demand for literacy to meet the more exacting needs of town life, but also provided the means of producing reading matter at lower cost than ever before.

Steam power made it possible to replace the old

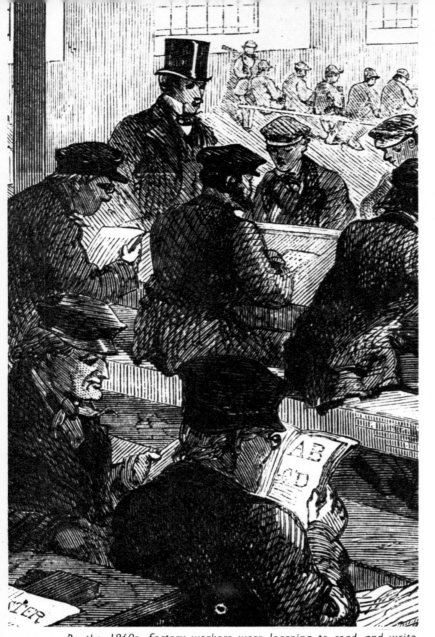
By the 1860s, factory workers were learning to read and write.
Mansell Collection

Steam power made it possible to replace the old hand-presses with faster rotary machines and so produce reading matter at lower cost.
Illustrated London News – Mansell Collection

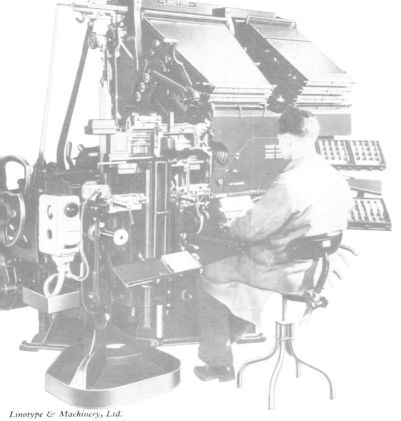

Linotype & Machinery, Ltd. / *Mansell Collection*

The invention of the Linotype and other type-setting machines meant that up-to-the-minute newspapers could be produced economically. By the 1890s printers and publishers of magazines were reaping profits by providing fare that looked both exciting and easy to read.

Mansell Collection

hand-presses by speedier machines with wheel-driven rollers. This enlisted a new process called *stereotype*. A hydraulic press gave the impress of the hand-set type to a moistened sheet of soft cardboard. By bending the dried card into the shape of a half cylinder and pouring molten type metal thereon, it was then possible to make a printing plate to fit the rollers.

Henceforth, it was practicable to produce sizeable up-to-date accounts of current events with commentaries on them. Yet for many years, only people of comfortable means could afford to buy a daily newspaper. Paper made from rags or rope-waste, though vastly cheaper than parchment, was still relatively expensive. The newspaper habit scarcely touched wage-earners till shortage of raw materials led paper manufacturers to use cheaper sources of cellulose: first grasses from North Africa, later wood-pulp from North America.

Even so, newspaper production was costly while it had to employ skilled craftsmen to assemble the type letter by letter. Two inventions in the latter half of the nineteenth century made the process more mechanical, and thus less time-consuming.

In 1835 posters were aimed at the few who could read with ease.
Alfred Dunhill, Ltd.

The Linotype and Monotype machines, in which a keyboard eliminates hand-setting, gave a wider public still cheaper reading matter.

Illustrated with pictures and advertisements for those who could read little if at all, the cheap newspaper stimulated the habit of reading and the demand for free schooling. Mechanical inventions which made printing less costly and more speedy were not the only circumstances which promoted the production of cheap newspapers. Trade rivals had begun to realise that it pays to advertise.

During the nineteenth century, the variety and number of goods manufactured by steam power mounted at a pace inconceivable during the first three centuries of printing. To sell so many new products in a competitive market, manufacturers and middlemen needed to make them known.

One way of doing this was to hire space in a newspaper. Thus advertising gave the newspaper proprietor a new source of revenue and press rivals the urge to increase their sales by selling newspapers at lower prices.

Advertising also fostered a new invention which has immensely improved the illustration of scientific text-books and the production of maps. When colour-printing as we now know it first became possible, enterprising businessmen began to hire space on walls or to set up screens on waste land by the roadside to draw attention to their wares by means of coloured posters. Politicians competing for working-class votes did the same.

Among pioneers of advertisement in Britain, still at the time leading the world in manufacture, Thomas Barratt paid in 1886 the sum of £2,000 for a picture painted by the artist Sir John Millais. He gave it the title of "Bubbles", and used reproductions of it to boost the sales of Pears Soap. Barratt himself founded Pears Cyclopaedia as a means of "combining practical information with general knowledge in one inexpensive volume". Its low price and attractive illustrations encouraged newly-literate artisans to begin buying and reading informative books.

Soon tobacco manufacturers and makers of breakfast foods were distributing with their wares cards with coloured pictures and a short botanical, zoological or historical text. Children who collected them not only became unintentional advertising agents; they also discovered that reading has a pay-off outside the schoolroom.

By 1880 mass production of cheap goods was in full swing. To sell their wares on the grand scale, manufacturers needed to attract the newly-literate factory hands. They advertised in the popular press, using eye-catching pictures and far fewer words.

Mansell Collection

Once colour printing got into its stride, cigarette manufacturers enclosed in their packets attractive cards with pictures and text on many aspects of history, botany, zoology.

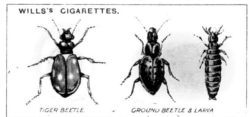

Carreras, Ltd.   *W.D. & H.O. Wills, Ltd.*

Société Rouennaise de Cirages
In this modern poster, bold colours arrest the eye and the slogan can be read at a glance.

This rotogravure machine prints four colours on each side of the paper in one operation and delivers folded sections ready for binding.
Terence Le Goubin

If the Prince of Master Printers, Benjamin Franklin, had not been a Founding Father of the first American Republic, he would have remained famous for his experiment on lightning; and if Newton had not provided the clue to how we now map the course of a man-made satellite, we should still remember him because of his experiments on the spectrum.

In one sense, we may say that his experiments pointed the way to colour-printing; but before cheap colour-printing was possible scientists had to discover more about how the eye matches a spectrum colour with the colour of a painted surface.

Newton showed that we can break down white light into a band of light of different colours, ranging from red at one end through yellow, green and blue to violet at the other end. He also showed that when we superimpose two such spectra in such a way that blue falls on yellow and red on green, we can again build up white light. When we say that red and green or blue and yellow are complementary, we thus mean that our eyes cannot distinguish a mixture of red and green light, or a mixture of blue and yellow light, from sunlight of the same intensity. We may thus recognise one surface as blue because it is painted with a pigment which absorbs all visible light except blue and another because it is painted with a pigment which absorbs only yellow.

But before the mid-nineteenth century, when men of science produced efficient spectrographs, it was not clear why mixing yellow and blue paints usually produces a green product. Today we know that this is because nearly all blue paints contain pigment which does not absorb green light unmasked by the neutralisation of blue and yellow. Knowledge of this kind now makes it possible to produce a wide colour range on the printed page by mixing, in various ways, inks of only three different colours, though some printers use black as well.

Using three photographic plates, each rendered sensitive to visible light of a different range of wavelengths, the printer can produce three printing plates, one for each of the so-called primary colours, red, yellow and blue. Each plate carries different intensities of ink on different parts of its surface.

The small pictures show some results of mixing different inks in different ways. Top: yellow only; middle: yellow and red; bottom: yellow, red and blue. Big picture shows result obtained from all four plates: yellow, red, blue, black.
John Markham

Percy Lund, Humphries Ltd., Bradford

The old Chinese script makes mechanical type-setting almost impossible. The reason is clear when we compare a Chinese and an alphabetic type-case. The contents of an alphabetic case can be reflected on a compact keyboard. A keyboard with one key for each of the several thousand signs in a Chinese case would be far too unwieldy.

Terence Le Goubin – L.T.A. Robinson, Ltd.

Not all parts of the world have been able to reap the full benefits of quick type-setting and cheap printing. Till lately, China, from which the Western World learned the use of paper and block-printing, has remained largely illiterate. The huge battery of meaning-signs which its traditional script employs makes a typewriter too bulky for office use and renders mechanical type-setting very costly. It also makes it hard to learn to read or to write. Thus modern China faces the future with 600,000,000 people, of whom less than 30 per cent can read.

We have already seen two reasons why the Chinese so long clung to their old script. First, languages spoken in various parts of China are widely different from one another, but scholars from every part of China can all read the same books in the old script. Next, because many sounds in Chinese languages convey many different meanings (like the single sound in the English words *you, ewe* and *yew*), there was reason to

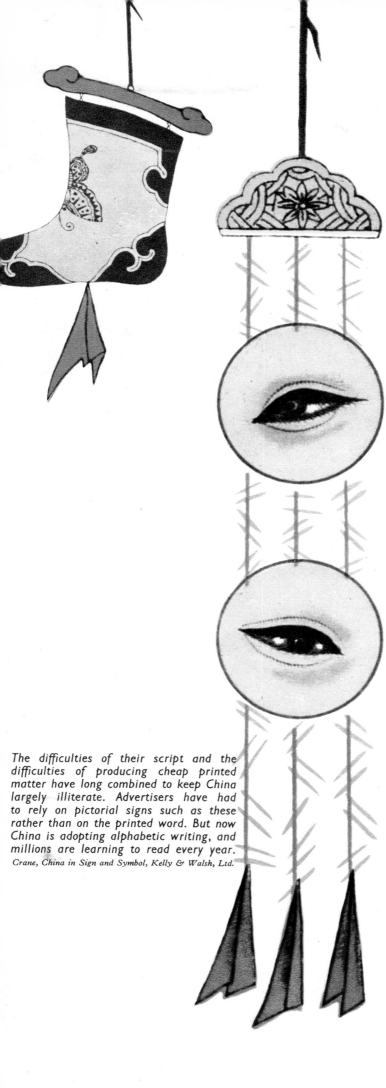

The difficulties of their script and the difficulties of producing cheap printed matter have long combined to keep China largely illiterate. Advertisers have had to rely on pictorial signs such as these rather than on the printed word. But now China is adopting alphabetic writing, and millions are learning to read every year.
Crane, China in Sign and Symbol, Kelly & Walsh, Ltd.

fear the danger of misunderstanding if they were to use alphabetic signs for their one-syllable words.

The second difficulty would also inconvenience us if we spelt *boy* or *buoy* in English as BOI, *son* or *sun* as SUNN and *read* or *red* as REDD; but we should still know whether the writer meant *son* or *sun* if he wrote the first as SUNN-BOI and the second as SUNN-REDD. In fact, the Chinese people use many such couplets when talking. Through a recent reform, they are now using an alphabetic script in schools and making use of the same trick in writing. They are also using accents to indicate tones, which play a large part in conveying the meaning of a one-syllable Chinese word.

People have always been slow to improve their means of communication. Indeed, big changes occur only after a big upheaval, as happened when a revolution and a war brought Kemal Ataturk to power in 1923. This enabled him to speed up Turkish education by substituting the Roman for the Arabic alphabet. Only since 1950 has it been possible to get agreement to adopt a new script in China.

Before then, China had no books in a script which conveys the sounds of words. Scholars of one part of China could learn the spoken language of another part only by ear, as children and parrots learn. Modern China is now dealing with this difficult problem in two stages. Children whose home languages are different will henceforth use books in alphabetic writing to help them to learn one common language (Mandarin). A generation hence it will be possible to print in alphabetic script books which people all over China will be able to read.

Turkish Embassy
Kemal Ataturk's script reform soon made Turkey a literate land.

To reap the benefits of scientific advances, literate workers are needed. In Ghana, children are now learning to read their own language.
Department of Social Welfare and Community Development, Ghana

If used for peaceful purposes, atomic power, which may otherwise destroy all human life, has offered mankind everywhere the possibility of release from irksome toil and hunger. Modern medicine has also given us the means of doubling the average length of life of people in tropical countries. Meanwhile, more than a third of the people of the world can neither read nor write; and until all their citizens are literate, many vast territories lack the labour force to take full advantage of the benefits science can confer.

In such territories UNESCO is now waging a vigorous campaign against illiteracy. Its success involves the solution of two different problems.

The first arises in communities which have no firmly established system of writing and in which few can read at all. Where this is true, as in Polynesia, the sound-pattern of the languages spoken may immensely simplify the task. Why this is so we have seen. When the words of a language are mostly words of two or more syllables, each syllable being either a simple vowel sound or a simple consonant followed by a vowel, the number of possible syllables is very small. It is then easy to represent each syllable by a Roman alphabetical sign, possibly with the addition of a few new characters. A simple grid on chart or blackboard then gives the learner a compact picture of every requirement.

For simplicity, suppose that such a language had only the four consonants represented by P, K, F, S and only three vowels represented by A, O, I. All the learner would need to do would be to learn to recognise the sounds represented by A, O, I; PA, PO, PI; KA, KO, KI; FA, FO, FI; SA, SO, SI. He or she would then be able to read anything whatever.

A scholarly American missionary, Dr. Laubach, devised this system for his letterless flock in a part of the Philippine Islands where the total number of syllables used in the native language is only 52.

Adults so taught can learn to read their own language proficiently in a couple of days if very quick, or in a couple of months if very slow.

To apply this method, there must be reading matter available in languages which have hitherto had little or no literature other than the sacred works of their missionaries. UNESCO encourages the production of cheap news-sheets which may pay for themselves by printing advertisements.

To be sure, many of these people will need also to learn English or some other language in which cheap books on difficult topics are obtainable. But experience shows that people who learn to speak English while learning to read their own languages, can later learn to read English more easily.

In Burma, Ceylon, Indo-China and parts of India there is a different problem. Buddhist teachers long ago introduced into these regions one or other sort of alphabetic writing which had already or has since become partly syllabic and even more difficult to read than the signatures of some poets or the prescriptions of most English physicians. Here the few scholars, like the mandarins of old China, have a vested interest in preserving scripts which slow down learning, make education more costly and are an obstacle to the use of mechanical type-setting.

This is mainly a political problem, which the people involved will have to solve for themselves with whatever encouragement the Western World can give them.

Department of Social Welfare and Community Development, Ghana
For older citizens, volunteers run special classes. Teachers and their pupils receive awards on Ghana's Mass Education National Literacy Day.

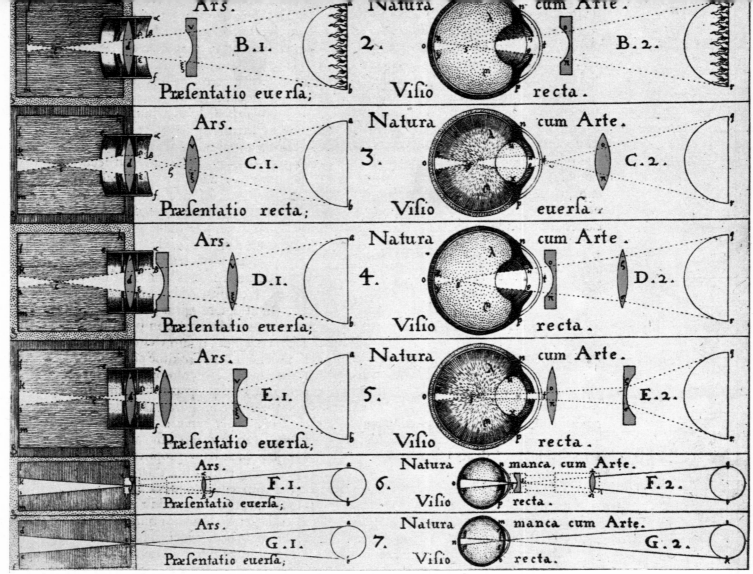

These diagrams of 1630 indicate real progress in optics. They show considerable knowledge of the properties of lenses and of the eye.
British Museum

Magic Lantern to Movie

During the three thousand years which elapsed between the coming of the alphabet and the beginning of printing from movable type, the art of painting pictures for the fun of it advanced little beyond the level of skill reached by the cave artists of twenty thousand years ago.

In Greece and among the Etruscans of Italy in the fifth and sixth centuries B.C. there had indeed been great progress in carving life-like solid models of people and animals; but no one had tried to make a flat picture look really like a solid object until about the beginning of the fifteenth century of our era. There were then wealthy Italian patrons prepared to buy realistic pictures from painters whose only means of livelihood had formerly been the decoration of churches.

The artists who first made pictures in perspective took the lead in studying images produced by light. What is more, they did so at a time when the science of optics was ready to take a big step forward for another reason.

In Italy, which led the way in making truly transparent glass suitable for lenses and windows of high quality, thirteenth-century monks had used lenses to help old people with bad sight. When printing brought reading matter within reach of many old people, more of them than ever before became conscious of the need for spectacles to correct long-sightedness. Thus a new industry of spectacle-making rapidly came into being.

In due course spectacle-making was to give rise to two powerful tools of discovery, the telescope

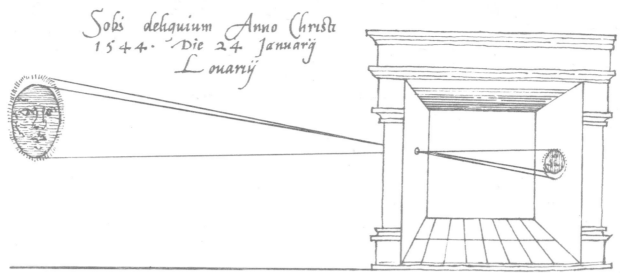

The camera obscura enabled 16th-century artists to focus an inverted image on a screen, then trace its outline.

and the microscope. Meanwhile painters of the sixteenth century began to use lenses to focus on a transparent screen an image from which they could trace an accurate outline of an object or a scene to form the basis of a picture. A device they employed, called the camera obscura, is still in use. Thenceforward, it was but a short step to the magic lantern, invented in the middle of the seventeenth century by the priest Athanasius Kircher.

All Kircher's lantern slides, depicting devils with fire-irons to remind evil-doers of the horrors of hell, were hand-painted. There was as yet no means of fixing a lens-made image on a transparent surface; but chemists already had a clue to how it might be possible to do so. By the seventeenth century, men of science of the Western World already knew of substances which darken on exposure to light, as do certain silver salts.

Photo. Science Museum, London

The 17th-century magic lantern was the first true projector. The slides were all hand-painted. There was yet no way of fixing an image.

In 1802, Thomas Wedgwood, son of the English Master Potter who was himself a man of high scientific attainments, used a lens to focus an image on paper coated with fresh silver nitrate not previously exposed to light. He succeeded in getting a black-and-white image which lasted for a brief period after withdrawal of the source of light. Needless to say, such pictures were quickly ruined unless kept permanently in the dark.

Further exposure to light blackened those parts of the silver compound which had remained unblackened by the original exposure. Wedgwood's experiments had therefore no practical outcome; but they stimulated others to seek for solvents which would dissolve the unblackened compound, leaving only the blackened part behind.

Two continental artists, Niepce who died in 1833 and Daguerre who survived till 1851, each of them familiar with the camera obscura as an aid to lithographic illustration, succeeded in this quest and so produced the first permanently fixed images; but their efforts resulted merely in producing direct positives from which it was not possible to make copies.

The inventor of photography as we know it today was Fox Talbot, an Englishman who used the camera obscura to make life-like sketches from nature. In 1839 he announced the details of his process to a meeting of the English Royal Society. His predecessors had required a fairly long period of exposure to produce a fixed image. Fox Talbot soon found that treatment with gallic acid could make silver salts blacken after a comparatively brief exposure to light.

Images produced by this method on glass covered with a film impregnated with silver nitrate and silver iodide were indeed negatives, and as such useful only to the artist; but Fox Talbot was quick to see that they could be used to print off any number of positives on paper impregnated with the same salts, developed and fixed in the same way.

The still photograph marks an immense enlargement of man's power to communicate knowledge. It makes it possible to produce likenesses beyond the skill of even the most accomplished draughtsman, to produce permanent records of momentary happenings, to make pictures of places inaccessible to the artist, and to bring a vast range of natural objects visually within the experience of people who would otherwise have no opportunities of seeing them.

To be sure, photography in its early days lacked colour; but the study of the spectrum first investigated by Newton had recaptured the curiosity of scientific men such as Helmholtz in Germany and Clerk-Maxwell in Britain. The possibility of producing coloured transparencies on a light-sensitive surface impregnated with particles of different pigments was not difficult to anticipate, because such men realised that any part of the surface would then blacken only if exposed to light of a particular range in the spectrum.

However, the technical problem of distributing the pigment particles so that any speck of visible size would contain all of them was a very tricky one, and there were many failures to devise a satisfactory process before colour-photography came into its own during the nineteen-thirties.

Gernsheim Collection
Fox Talbot's success in producing positives from negatives made possible photography as we know it. This is his photographic establishment.

Daguerreotypes were used chiefly for portraits.
By courtesy of the Director, Science Museum, London

The still camera can record momentary events like the Hindenburg disaster of 1937.
Sam Shere – Planet News Ltd.

It can picture a nebula many thousands of light-years away ...
By courtesy of the Mount Wilson and Palomar Observatories

... and reveal the complex structure of the crystals of vitamins.
Hans Waldmann

It can produce a picture of life in the world's great oceans ...
Hans Hass

... or show us how the earth looks from a rocket 140 miles away.
Official U.S. Navy photograph

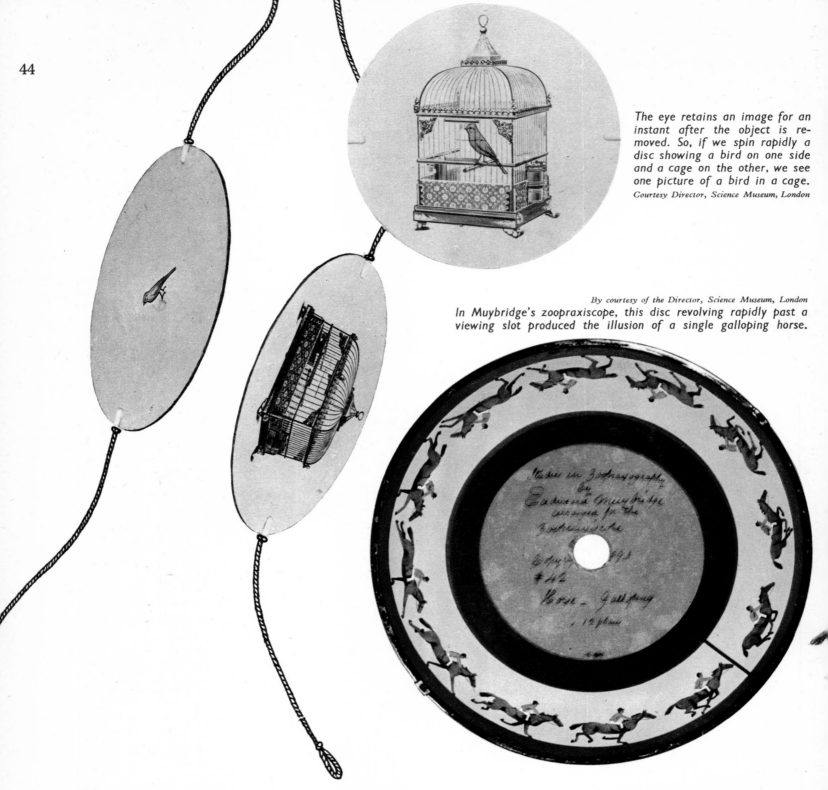

The eye retains an image for an instant after the object is removed. So, if we spin rapidly a disc showing a bird on one side and a cage on the other, we see one picture of a bird in a cage.
Courtesy Director, Science Museum, London

By courtesy of the Director, Science Museum, London
In Muybridge's zoopraxiscope, this disc revolving rapidly past a viewing slot produced the illusion of a single galloping horse.

In the closing years of the eighteenth century, craftsmen found better methods of polishing glass and of making lenses which could magnify greatly without creating a colour-fringe to blur the outline of the image. This led to the construction of better telescopes and microscopes. It led also to a keener interest in the study of light during the early years of the nineteenth century, when photography was in the experimental stage.

Those who studied light had long been accustomed to regard the eye as a camera obscura. They now made the discovery that it is like the photographic camera in one way. That is to say, the image on the sensitive layer called the retina persists for a fraction of a second after withdrawal of the source of light.

The consequence of this is that images presented to the eye in rapid succession blend into a single image which seems to be in motion. An early use for this discovery was to make children's toys called by various long names. Such devices produced moving images by the spin of a cylinder or a disc carrying successive pictures of the same object in different stages of motion.

In the mid-nineteenth century moving images of a sort were on show in a new setting. The electric arc lamp had now made it possible to cast bright images on a distant screen, and the introduction of photographic slides suddenly made the magic lantern far more popular than ever before as a means of showing still pictures of life in foreign lands to large audiences. It was not hard to adapt it to show a brief moving picture by passing several slides in quick succession between the lens and the lamp. But it was still difficult to foresee the possibility of keeping the image in motion for more than a second or so.

One ingredient of success was still lacking. Hitherto the only highly transparent substances known to mankind were hard, brittle solids such as glass, quartz and fluospar. To show a vast number of successive images blending to make a moving picture lasting more than a few seconds, the magic lantern showman needed a transparent material sufficiently flexible to unwind quickly from a reel.

Industrial chemists, who first produced such a substance in the eighteen-sixties, called it celluloid. For photography it was better than glass, and for much the same reasons that papyrus is better for writing than are clay tablets. It is lighter, less liable to break, and much easier to store in a small space. It is convenient to the photographer because a single compact reel dispenses with the need for changing the plate before taking another snap.

The great American inventor, Thomas Alva Edison, was not slow to see how such a reel of celluloid tape, perforated at its edges to fit into the cogs of a wheel, could produce a continuously moving picture when focused on a screen by the magic lantern. Such was the beginning of moving pictures in the modern sense.

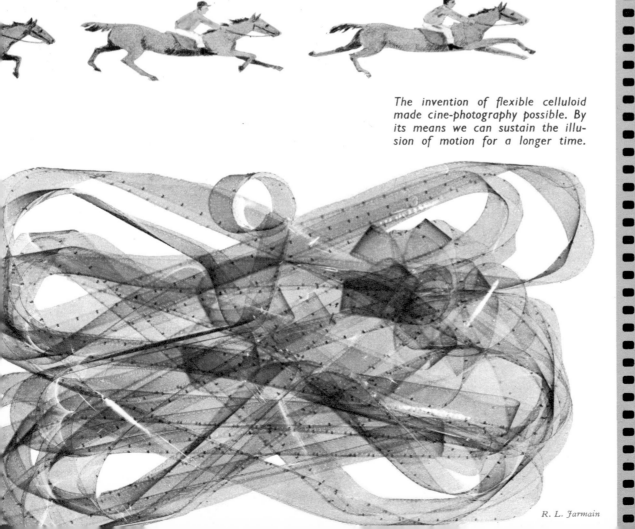

The invention of flexible celluloid made cine-photography possible. By its means we can sustain the illusion of motion for a longer time.

R. L. Jarmain

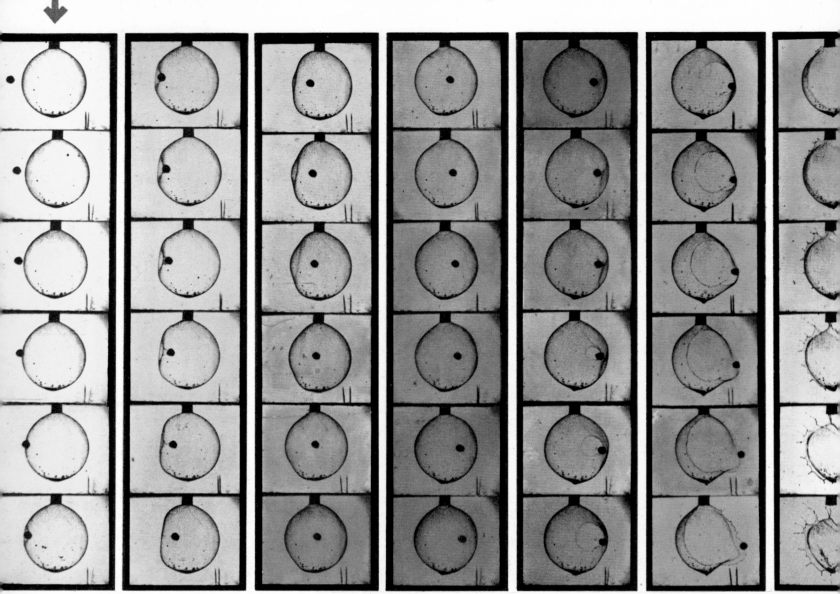

The film can show movement in a new way. Screened less rapidly than photographed, these shots show what happens when bullet hits bubble.

By courtesy of the Director, Science Museum, London

The cinematic film is immensely more powerful both as a means of entertainment and as a means of spreading knowledge than its parent, the magic lantern. It is also a research tool, because it enables us to study growth and movement in a new way. We can use it to speed up our view of a slow process, such as the growth of a bean tendril, or to slow down the representation of a quick one, such as the golfer's stroke or the impact of a raindrop striking the ground. With it, we can build up libraries of visually-recorded happenings of historical interest; we can also use it to speed up education, especially where there is a shortage of teachers.

The discovery of a transparent plastic material made the cinematic film a reality only half a century after such a possibility was first thinkable. Commercial production of other plastics played an almost equally important part in promoting another invention of the second half of the nineteenth century – the phonograph, which made it possible to record, store and reproduce on demand both human speech and music. The marriage of the two inventions about 1925, when the talkies first became a commercial enterprise, was inevitable. The combination raised no very new mechanical difficulties, but it brought the maker of films face to face with a very old human problem.

The early film made its impact by pictures alone; it could appeal to people of all lands, regardless of language differences. The talkie needs a different sound track to make it suitable for each different speech-community. It will be possible to make the fullest and most economical use of the modern film only when people who use different languages in their homes learn to use one and the same tongue for communication across language frontiers. Only then will it also be possible to take the fullest advantage of the opportunities which radio offers to enable men and women of all countries of the world to communicate instantaneously, as if face to face.

A compromise between still and cinematic photography can provide useful lessons for the sportsman.
H. E. Edgerton

By kind permission of the Rank Organisation, producers of A Tale of Two Cities.
The irregular zigzag line to the right of the pictures on this film is the sound track. The poster below emphasises how sound tracks bring the maker of films face to face with the language problem.

Canadian National Film Board

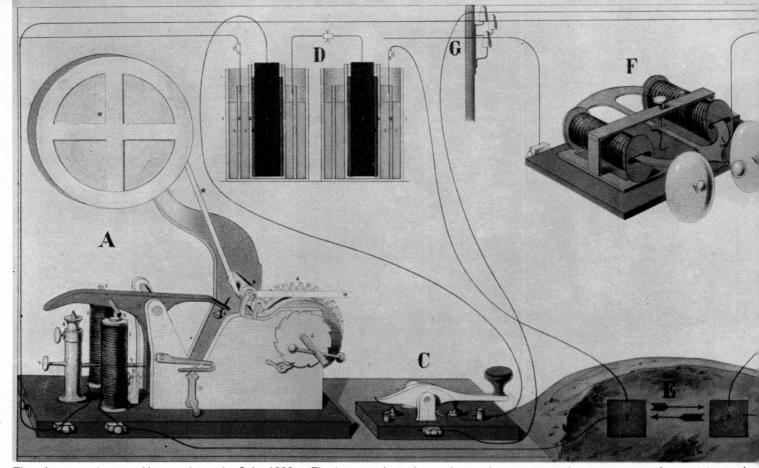

This diagram shows a Morse telegraph of the 1880s. The key-switch at the sending end operates an electromagnet at the receiving end.

Telegraph to Television

Our story has shown us how the only talking and picture-making animal on our planet has learned to convey messages beyond the range of the voice and beyond the grave: first by painting on cave walls, then by writing signs which were essentially pictorial, later by writing signs which recall the sounds we utter in speech, later still by engravings and etchings of which the printer can make many copies, and last of all by means of photography.

From time immemorial, people had communicated at close range in other ways – by the beat of drums, by beacons on hilltops and by coded flag signals. The study of the electric discharge in the eighteenth century suggested the possibility of communicating at a greater distance by using a current. Early in the nineteenth century two events made the idea practicable: the invention of new generators and the discovery that a current can be used to deflect a nearby magnetic needle or to magnetise a piece of iron. Within a few years the coming of railways and fast-moving locomotives made electrical signalling indispensable.

The first railway telegraphs made use of the fact that reversal of current changes the direction – left to right or *vice versa* – of a magnetic needle in its neighbourhood. They had thus to use a code of only two signs – right and left. Such a code can accommodate 2 arrangements (R or L) for one release of current, 4 arrangements (RR, RL, LR or LL) for two releases, 8 for three, 16 for four, and so on. Thus a 5-stroke code can convey 32 different signs – enough to represent both a modified alphabet deprived of useless signs like Q or X, and the essential numerals 2 to 9.

The later Morse telegraphs used the fact that the current in a coil round a piece of iron will attract an iron lever audibly; but the 2-sign code (a short or a long percussion of the lever, which can be printed on a moving tape as dot or dash) had come to stay.

These new codes remind us of a secret form of writing called Ogham, used by the Celtic peoples of Ireland, Wales and Scotland about the time when the Roman legions withdrew from Britain.

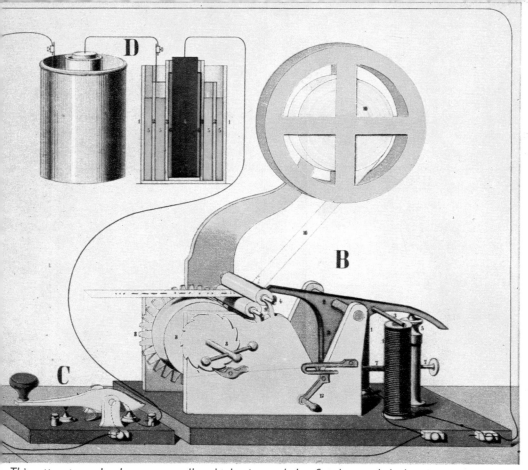
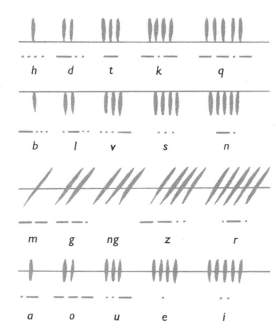

The Morse Code uses only two signs, dot and dash. Ogham script uses vertical and oblique strokes.

This attracts and releases a needle which pierces holes for dots and dashes in a moving tape.

This photograph reminds us that it was the unprecedented speed of the early locomotives that first made electrical signalling indispensable.

John Gay

We can now decipher them from stone monuments which record the Latin or Gaelic equivalent; but we do not yet know who invented them – or why.

To send a message along the telegraph wires, it was necessary to spell out each word in a code employing several repetitions of two signs for each letter. So telegraphy is a slower way of communicating than speech, and, like the earliest alphabetic writing, is chiefly useful for sending short messages.

People had long known that different sounds cause a nearby metal diaphragm to vibrate in different ways. Probably it occurred to several inventors that it might be possible to use the vibrations of one diaphragm, resonating to the human voice, to vary the current in the coil of a magnet in contact with a second diaphragm, which would then vibrate in unison with the first. In 1876 an American inventor named Bell first succeeded in transmitting speech in this way.

The telephone, as we call his invention, made conversations at a distance far beyond the range of the human voice possible for the first time. Large trading firms were eager to place orders more quickly by taking advantage of it; and its use spread rapidly, especially after 1900, when electric power became cheaper. Over the years, the laying of telephone cables across the continents and along the beds of oceans, has made it possible to carry on a conversation between almost any two parts of the earth. While Bell was completing his invention, another American inventor succeeded in recording the human voice mechanically.

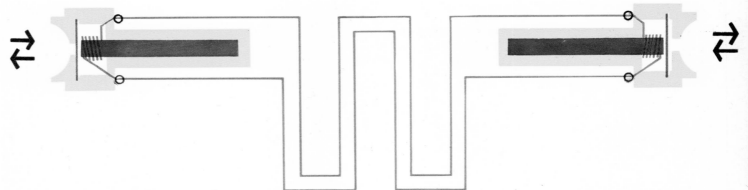

In the early telephone a voice vibrated a diaphragm, switching on and off a current controlling an electromagnet in the receiver. The current from this electromagnet caused a second diaphragm to vibrate in time with the first, reproducing the words of the speaker.

Bell Telephone System
Direct overseas telephone circuits to and from the U.S.A. Today it is possible to hold a conversation between any two parts of the earth.

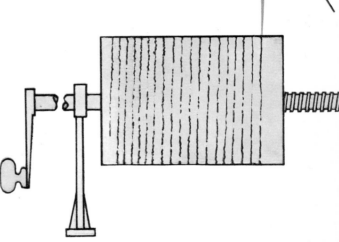

The great actress, Sarah Bernhardt, listening to a recording of her voice. In the early phonograph, a voice set a needle vibrating, cutting grooves in a rotating waxed cylinder. At play-back, the grooves set up similar vibrations in another diaphragm.

Terence Le Goubin – B.B.C.
Libraries of tape-recordings and long-playing discs may be vital reference sources for future historians.

Edison's first phonograph transmitted vibrations set up in a diaphragm by the human voice to a cutting edge scratching on a rotating wax cylinder. The cylinder, now carrying an irregular spiral groove on its surface, was set rotating again with a soft needle following the irregularities of the groove. The vibrations of the needle, transmitted to a diaphragm, then reproduced the speaker's words.

Modern record-players are essentially like the early phonographs; but the disc has replaced the cylinder record because it is more easy to store. Thus we can now hear the words of the dead as well as of the living, and the music of orchestras long since disbanded as well as those of today. We can have, and indeed broadcasting corporations do have, libraries of sound. Together with films they are becoming an increasingly important part of man's artificial memory.

We know what Napoleon looked like only from hand-painted, and possibly flattering, portraits, but we can never recapture his voice and mannerisms. The cinema and the phonograph bring recent history more vividly before us. They give us a permanent record of what Hitler looked like, of his gestures and of the way in which he spoke.

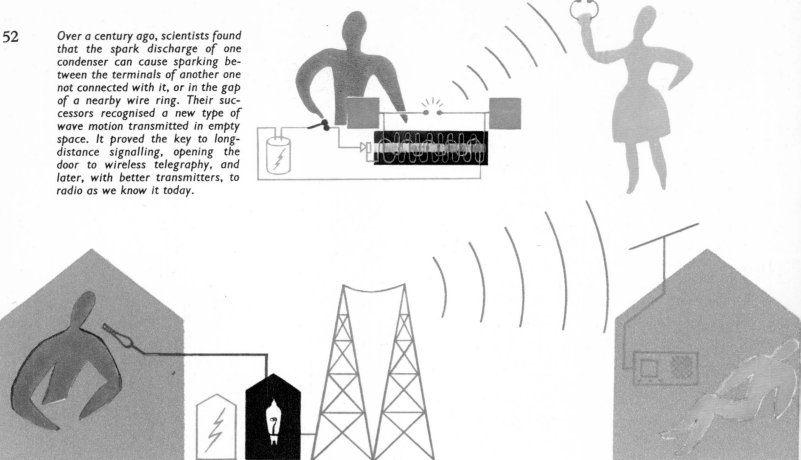

52 Over a century ago, scientists found that the spark discharge of one condenser can cause sparking between the terminals of another one not connected with it, or in the gap of a nearby wire ring. Their successors recognised a new type of wave motion transmitted in empty space. It proved the key to long-distance signalling, opening the door to wireless telegraphy, and later, with better transmitters, to radio as we know it today.

Radio Times Hulton Picture Library

It is now an everyday occurrence for many of us to hear a crackle in the radio set produced by the sparking of a nearby motor. A hundred years ago, the discovery that the spark discharge of one condenser can cause sparking between the terminals of another, suitably adjusted but not connected with it, was a momentously novel event. The study of this phenomenon led to the recognition of a new type of wave motion, transmitted in empty space like light and radiant heat, but, unlike the visible radiation from the beacon on the hilltop, not readily absorbed by material objects.

Here then, was the master-key to signalling over great distances without the cost and inconvenience of laying cables or erecting telegraph poles. Before the first decade of our century ended, an Italian inventor, Marconi, made use of the new discoveries to send long-distance signals without wires. By 1912 wireless telegraphy linked continents separated by oceans, and made it possible for the first time for ships at sea to receive signals from port or to send warnings to other vessels afloat. Because it employed a dot-and-dash code, the wireless telegraph was mainly useful for short messages.

Other inventors were soon on the track of a suitable transmitter to resonate the discharge in unison with the voice; and the discovery that the outer layers of the atmosphere reflect short-wave electromagnetic radiations back to earth convinced them that sound transmission from any part of the world to any other part at a speed equal to that of light was no idle dream.

The receivers of the first cat's-whisker radio sets available at the end of World War I were earphones of the sort still used for hospital patients. Only one person could then listen in at the same set; but within a few years it was possible to amplify the sound so that an audience could listen without such hearing aids.

Before the beginning of World War II, this brought into homes of people with modest means a new form of entertainment, depending wholly on sound to achieve its effects. It also provided governments and powerful political parties with a new way of influencing mass opinion, vastly more effective than the printed newspaper.

From a mechanical viewpoint, sound broadcasting had a world audience; but the possibility of transmitting a universally understandable message from a single human voice was not yet in sight. People in places as far distant as Auckland and Aberdeen, Washington, D.C. and Wigan, England, or Melbourne and Manchester, can all understand the same broadcasts; but most listeners in places as close together as Canterbury and Calais, Bordeaux and Barcelona, the Hague and Harwich or Mexico City and Missouri are not able to do so. They speak different languages.

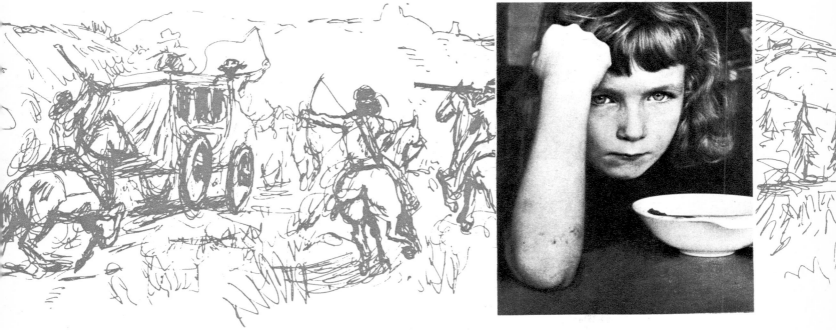

Joan Miller

Before the advent of television, radio had to depend wholly on sound to achieve its effects. From a mechanical viewpoint there was no reason why it should not have a world-wide audience. The language barrier, however, confined its audience to a single speech community.

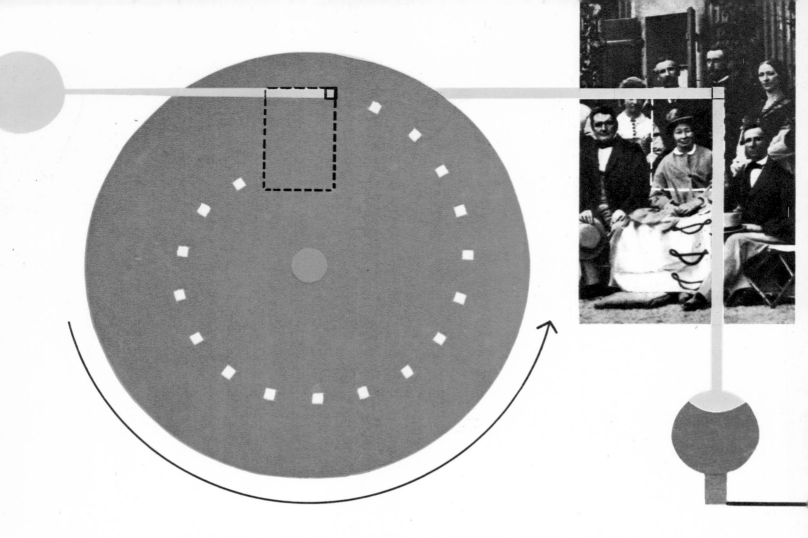

The television scanning principle is based on Nipkov's disc. A beam of light projected through each hole as the disc turns traces a series of lines across a picture in front of the disc. If the disc turns fast enough, the eye sees a solid picture. With two synchronised discs and a device which conducts a current only when light falls on it, it is thus possible to reproduce a picture on a screen in front of the second disc.

We have seen that the cinema is the offspring of a marriage between two inventions, one of which depends on the persistence of an image on the retina. So is television.

Among other devices which employ this principle is a disc first constructed by a German inventor, Nipkov, in 1884. Nipkov's disc has fine holes arranged in a spiral. When it is set spinning, each hole in turn lets through a fine beam of light from an illuminated screen behind. If it rotates fast enough, the separate images the eye receives from the individual holes appear to fuse in a continuous pattern of light and dark, changing as the pattern on the screen changes.

If two such discs rotate at the same speed, the second will transmit the same image from the same source of light. To exploit this for transmitting moving pictures at a distance, however, two more ingredients are necessary. At the transmitting end there must be a screen which responds to light by creating a current; at the receiving end there must be a source of light which comes into action without the time-lag of an ordinary electric lamp.

Discoveries made in the 1870s led to the invention of the photo-electric cell, which met the first need. Since this device conducts a current only while light falls on it, it equipped the transmitter with a light-sensitive switch. Neon-tube lamps, first available for commercial use about 1920, provided a satisfactory source of light for the receiver.

In 1926, a Scottish engineer named Baird gave the first demonstration of how moving images can be transmitted by radio waves at a distance. The images were of sufficient precision to hold out the prospect of success. In little more than twenty years television had become a powerful new means of mass communication.

Unlike sound broadcasting, television speaks the universal language of the cave paintings; but language barriers still defeat the full realisation of global communication by simultaneous appeal alike to the ear and to the eye.

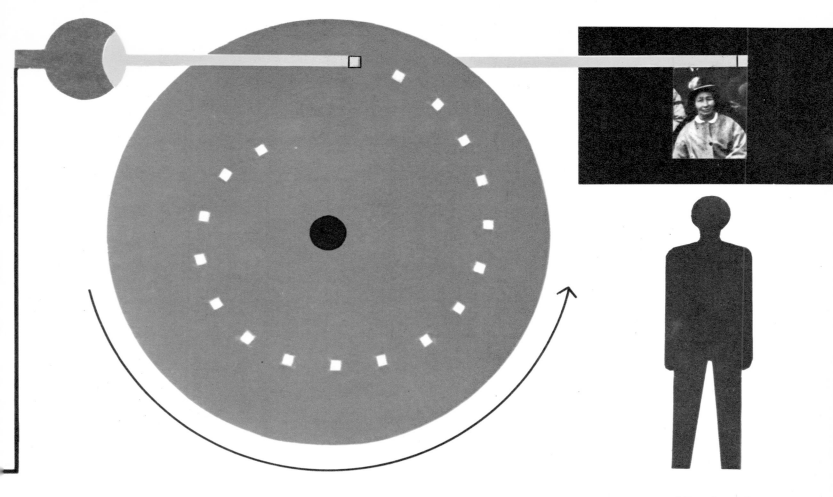

At the Coronation of Elizabeth II, there were less than 8000 people in Westminster Abbey. But millions saw the ceremony on television.

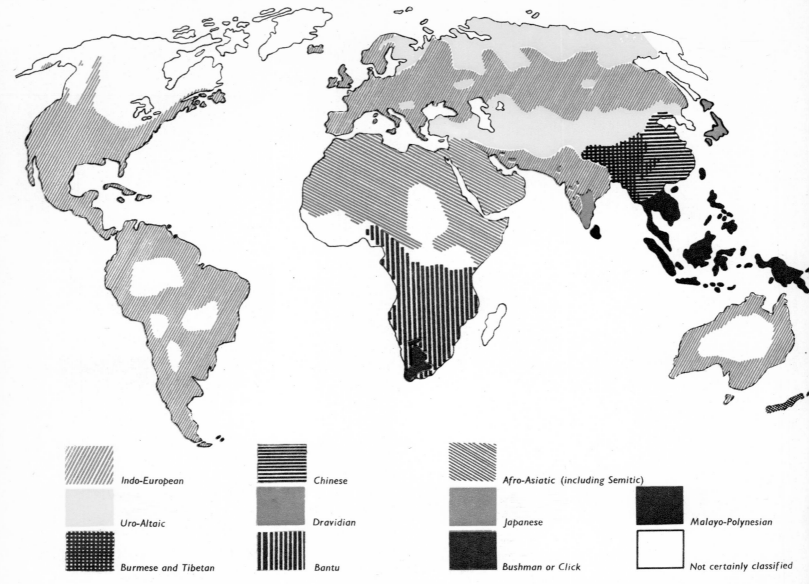

This map shows the main language families some, though not all, scholars recognise. Some make Burmese – Tibetan and Chinese one group.

Our One World

Within a country where all people can communicate in what we customarily call a language, there may be many local varieties of speech, distinguished by differences of pronunciation and words in common use. There is no clear-cut division between languages and such local varieties which we call dialects, because speech changes in different ways in different places, and dialects become more different as time goes on. For instance, people who spoke Latin dialects of Gaul, Italy, Roumania and the Spanish peninsula could talk freely together in A.D. 400. Today they cannot do so; but it is easy to see that many words of French, Roumanian, Italian and Spanish as spoken today are alike. Thus we regard all these languages as members of one family with one ancestor.

Similarly, it is easy to see that Dutch, German, Old English, Swedish, Danish and Norwegian have the same sort of family likeness. Very often, such a family likeness is beyond recognition unless we make allowance for consistent changes of pronunciation in different regions of a territory which uses one language. For instance, we have already seen that in some regions the *s* sound consistently replaces the *h* sound; many Latin words are like many Greek words if we substitute the *s* sound of *solar*, *saline*, *septuagenarian*, *semichord* and *sexagesimal* for the *h* sound in *helium*, *halogen*, *heptagon*, *hemisphere* and *hexameter*.

Another way in which language changes is by slurring over the middle or lopping off the end of words in careless speech. Thus we recognise more

Da nobis hodie panem nostrum quotidianum.

Gib uns heute unser täglich Brot.

Donne-nous aujourd'hui notre pain quotidien.

Geef ons heden ons dagelijksch brood.

Danos hoy nuestro pan cotidiano.

Giv oss i dag vårt dagliga bröd.

'Give us this day our daily bread'. If we translate this line of the Lord's Prayer into (reading downwards) Latin, German, French, Dutch, Spanish, and Swedish, we can see some features which distinguish Romance languages (blue) from those of the Teutonic group (red).

confidently that the modern English word *head* is equivalent to Danish *hoved* when we are able to trace it back to its Old English ancestor *hēafod*.

So scholars can group dialects as languages, languages in small families, and such small families in larger ones. The Latin family, the Dutch-Swedish family (including English in its earliest form) together with Russian, Persian and most of the languages of North India and Pakistan make up the largest of all, the Indo-European family.

Our first map shows the territory of several other large language families: the Semitic, including Arabic, Hebrew and Maltese; the Chinese and the closely-related Burmese-Tibetan; the Uro-Altaic, which includes the Finno-Ugrian languages of Finland and Hungary; Japanese; the Malayo-Polynesian which includes Fijian and Maori; the Bantu which includes Swahili and most of the languages spoken in Africa south of the Equator.

Today there are translations of at least part of the Christian Bible in about a thousand languages. The number of languages spoken in the world is probably more than twice as great; and about many of them we know too little to recognise a common family likeness to any others. Today more people speak English than any other. Probably Russian and Spanish come next.

Adapted from Maillet et Cohen, Les Langues du Monde, Centre National de la Recherche Scientifique

The Indo-European family. Note that the languages of Sweden and Finland have less in common than those of Spain and Northern India.

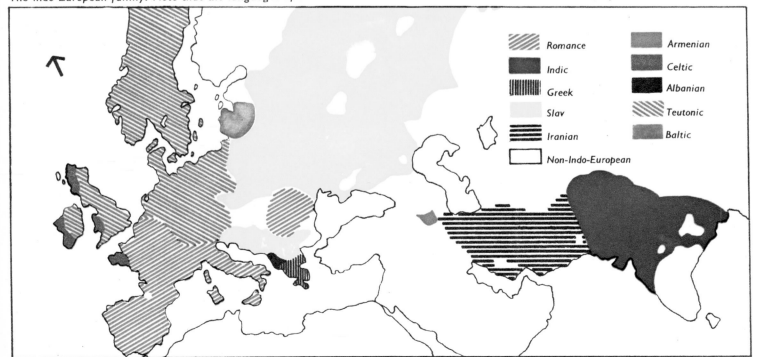

In these days of international travel it is a matter of survival that motorists everywhere should be able to understand traffic signs.

When printing began, men of science in the Western World used Latin, Hebrew or Arabic textbooks and wrote in one or other of these languages. In the seventeenth century, investigators and inventors began more and more to write in their own languages; and it became more and more difficult for those living in one part of Europe to learn about discoveries in another.

In the time of Newton, there were several unsuccessful attempts to get agreement about the use of a single language for science; and the hope of finding a remedy for the confusion did not revive till two hundred years later. Meanwhile, however, expanding scientific knowledge created the need for piecemeal reforms which took effect before the middle of the nineteenth century.

Before A.D. 1800, authors of works on the classification of plants were beginning to use standard conventions to convey the structure of different types of flowers. During the nineteenth century, electrical engineers developed their own symbols for circuits, chemists for the structure of the molecule of carbon compounds, and physiologists for the tracts within the nervous system. Today science has therefore an immense battery of picture-signs.

In Newton's time, biologists and chemists used local names for plants, animals and minerals. Such names, which were not always meaningful to people from different parts of a single territory with a common language, convey nothing about the family likeness of creatures or the composition of substances. During the latter part of the eighteenth century, and chiefly through the influence of Linnaeus, a Swede, and of Lavoisier, a Frenchman, men of science began to name things in a new way. Building up words from bricks which are themselves

Greek or Latin words, they produced new and far more informative names.

For instance, what the English had called spirits of salts now became hydrochloric acid. The word *hydrochloric* reminds us that the acid is a compound of hydrogen and chlorine. The latter, a green gaseous element, gets its name from the Greek word for green (as in *chlor*ophyll) and hydrogen gets its name from two Greek words meaning water (as in *hydr*ant) and forming (as in *gen*esis).

This new way of naming things aims at packing as much information as possible into the name. Thus a few rules tell us that the name nitrogen trioxide is a label for a compound of which we can make two litres by combining two litres of nitrogen with three of oxygen. Similarly the name Equus caballus for *horse* (English), *cheval* (French), tells us that it is a close relative of Equus asinus, i.e. *ass* (English), *âne* (French), placed with it in the same zoological family, Equidae.

So it has come about that scientific men of all nations now have a vast collection of names which all of them can understand. Because science constantly enriches our daily lives, many of these names are now in daily use everywhere. In such words as *geophysical*, *aeroplane*, *pyrex* and *dehydrated*, we now, if unwittingly, rely on the words used by Aristotle for earth, air, fire and water.

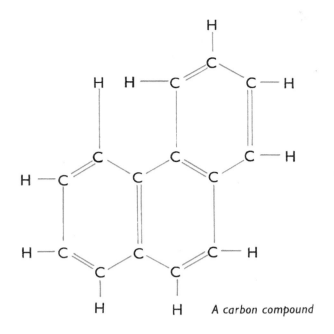

A carbon compound

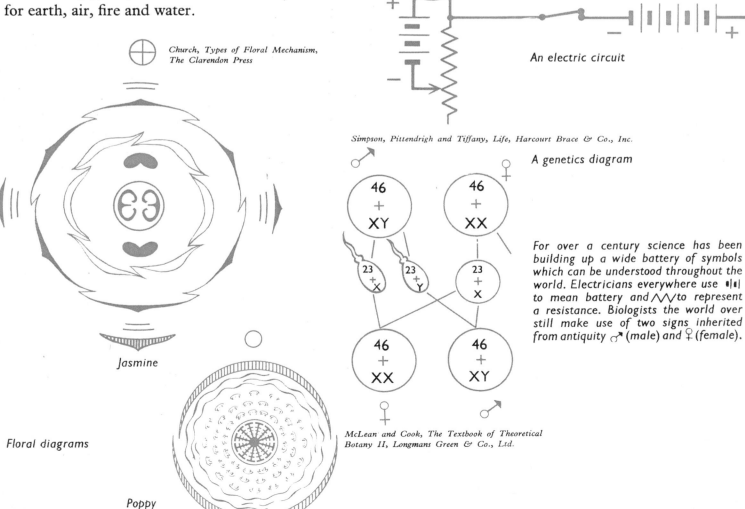

An electric circuit

A genetics diagram

For over a century science has been building up a wide battery of symbols which can be understood throughout the world. Electricians everywhere use ⊣|⊢ to mean battery and ⋀⋁ to represent a resistance. Biologists the world over still make use of two signs inherited from antiquity ♂ (male) and ♀ (female).

Floral diagrams

Jasmine

Poppy

Church, Types of Floral Mechanism, The Clarendon Press

Simpson, Pittendrigh and Tiffany, Life, Harcourt Brace & Co., Inc.

McLean and Cook, The Textbook of Theoretical Botany II, Longmans Green & Co., Ltd.

60 Although inventors and investigators still have to rely largely on languages they cannot all hope to share the use of, they have a large common battery of picture-signs, proper names, and mathematical symbols. What is perhaps more remarkable, they all use the same units of measurement.

This is one of the great cultural benefits bestowed by the French Revolution. In 1791, Lavoisier, who was foremost as a promoter of the new way of naming substances, became secretary and treasurer

Service photographique – Bibliothèque Nationale, Paris
Before the National Assembly made the use of the metric system compulsory in France, apothecaries, goldsmiths and bakers all used different standards of weight as they still do in Britain. Now many parts of Europe and of South America employ the metric system for all purposes. In the laboratory its use is universal.

of a commission set up by the National Assembly to revise the French system of weights and measures. At that time, as is still true in Britain, different trades had different standards of mass and volume, and they had no consistent plan of dividing or multiplying units to make smaller or larger ones. The revolutionary commission adopted a single standard of length (metre), simply related to a single standard of mass (gram), and defined all subsidiary units, e.g. centimetres and kilograms, as decimal fractions or multiples of these two.

By the time the commission completed its work, a new estimate of the great circle between Dunkirk and Barcelona, based on eight years' research, made it possible to define a quarter of the earth's circumference with much greater precision than ever before. The commission agreed to define the metre as one ten-millionth of this distance. The gram is the mass of one millionth part of a cubic metre of water at the temperature (4° on the Centigrade scale) at which its density is greatest.

These definitions lead to a convenient unit of heat (calorie) based on the amount required to raise one gram of water through one degree on the Centigrade scale. They also lead to a straightforward unit of force (dyne) – the amount which, acting on one gram for one second, produces a velocity of one centimetre per second. Within a century, scientists in all countries recognised the convenience of the new system and adopted it for use.

It is of interest that the system relies on the same battery of Greek or Latin numerals for all secondary units: *milli* = 1/1000, *centi* = 1/100, *deci* = 1/10, *deca* = 10, *hecto* = 100, *kilo* = 1000. The child who learns it is also learning how to use decimal fractions.

Though it avoids the waste of school time involved in memorising many different tables, some people were slow to recognise its usefulness. It was compulsory throughout France from 1807 onwards; but thirty years later it was necessary to impose heavy penalties to enforce its use. By then, other countries were beginning to recognise its advantages.

The Netherlands and Greece had decided to adopt it by 1840. Germany followed their lead in 1868; and in 1875 twenty nations sent delegates to France to take part in defining a convenient yardstick of the new measurements. To be sure, Britain still lags behind; but nearly all countries of Europe and South America now use the C.G.S. (centimetre-gram-second) system, based on the metric system.

CARL A. RUDISILL LIBRARY
LENOIR RHYNE COLLEGE

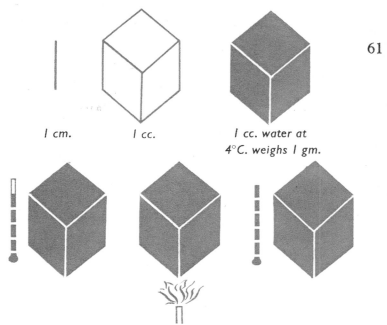

Heat needed to raise temp. of 1 gm. water by 1°C. is 1 calorie. Metric units of length, volume, weight and heat are all related.

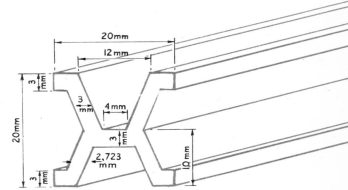

The International Prototype Bar is designed in such a way that it defines not only the metre, but smaller measurements as well.
Photo – Science Museum, London

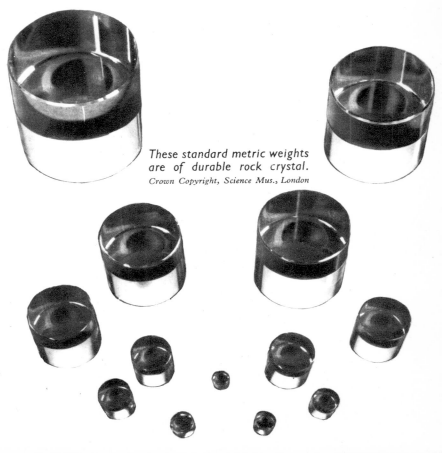

These standard metric weights are of durable rock crystal.
Crown Copyright, Science Mus., London

When men of science in Newton's time began to worry about the lack of a common language which discoverers and inventors from different European countries could all use, few other people felt the need for one. Travel was still tedious, costly and time-consuming, and nobody could foresee how scientific inventions would multiply opportunities for exchange of goods and of information between folks living far apart.

Conferences of engineers, medical experts, trade unionists, authors, salesmen and even students, attended by delegates from nations of all five continents, are now a daily happening. Before the Steam Age there were no such gatherings, even for all European nations; and the only countries contributing greatly to scientific discoveries were countries where people spoke English, German or French.

Today science is rapidly advancing in the Soviet Union, in Japan, in China, in India and in Pakistan. Tomorrow, Africa and South America may be making big contributions. Every year scientific journals published in scores of different languages pour out important announcements of new knowledge; but only a small proportion of the scientists and technicians who could profit by reading them can do so.

Meanwhile, international conferences called to settle disputes between nations often fail to do so, partly because those who take part can make contact only through interpreters liable to err about their intentions; and international machinery to promote goodwill has to allocate colossal sums on translation which delays the publication of arguments, proposals and agreements.

Madeleine Rands

The Brussels Exhibition was unique, but international meetings, as such, are commonplace. They underline the need for a common language.

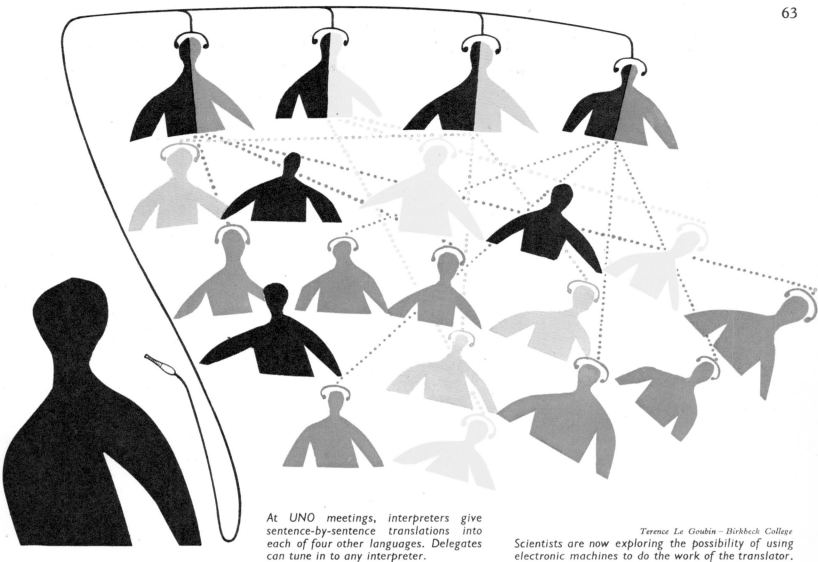

At UNO meetings, interpreters give sentence-by-sentence translations into each of four other languages. Delegates can tune in to any interpreter.

Terence Le Goubin – Birkbeck College
Scientists are now exploring the possibility of using electronic machines to do the work of the translator.

By the second half of the nineteenth century, such a state of affairs was already foreseeable to men of vision in several different countries, and there were already attentive ears to proposals for providing everyone with one and the same second language. Those who believed that this is possible could point to two facts.

One is that schoolchildren in most countries already learn some second language, such as English, French, German or Spanish. The other is that people in countries where citizens use different languages in the home are bilingual. For instance, Welsh is the language of the Welsh home; but all Welsh schoolchildren learn English which they use when they talk to their neighbours across the border.

On the face of it, there is therefore nothing unreasonable in the hope that all nations will eventually agree to give instruction to schoolchildren in one and the same second language. Twenty years ago, this seemed to be the only way of breaking down language barriers.

Today the governments of the United States and of the Soviet Union, as well as private citizens in Britain, are spending lavishly on projects to explore the possibility of making machines which can do the job of the interpreter. The fact that governments are willing to encourage these endeavours shows that the hope of breaking

This loom is controlled by holes punched in a series of cards. Textile mills have used such machines for nearly fifty years.
The Property of Bianchini, Ferier Co., Ltd. - I.C.T. Ltd.

through the sound barriers of speech is no longer the pipe dream of a few visionaries. Though it is still far too early to prophesy that a machine will ever do the job of the interpreter or of the translator, we may be sure that the attempt to make one will teach us much more than we understand as yet about the use of language.

The many inventions which have gone to the making of machines capable of translating simple messages from one language to another illustrate how largely machinery has displaced skilled clerical work during the lifetime of people now living. For more than half a century we have had recorders which translate the telegraphic code from Morse to Roman – or other – writing-signs. We have also had machines for carrying out ordinary calculations and typewriters which dispense with the need for the pen. These machines have no memories; but several devices more recently perfected store information for use when required, as our brains do.

One of these, the tape recorder, is merely an adaptation of the phonograph, and its memory is at best like that of a parrot. A second type, the mechanical sorter, can answer questions of a kind, if we supply it with information of the kind that it can take in. The earliest sorters came into use for census work, accountancy, and large-scale betting firms. For such purposes, all the required items of information are numbers referring either to sums of money or to populations.

Such machines cannot take in words; but they can take in numbers. If we lay out a card in columns labelled 1 to 80, each of ten rows labelled 0 to 9, we can represent any number less than one-followed-by-eighty-zeros by means of holes punched in row-cells of successive columns, as in our picture. If a conveyor passes such punched cards over a plate studded with pins whose release can operate a switch, it is possible to arrange a magnetic device which will trap in a particular box all those cards which bear a particular sequence of holes. By another fitment, it is then possible to count all cards which have the same pattern of holes. Simple acts of translation on machines called sorters and tabulators are now an everyday occurrence in large hospitals and in business houses.

Suppose that a physician wishes to know the names of all married women he has treated for diabetes. Let us assume that every patient has a code number of, say, 6 figures recorded on the first six columns of the punch cards, that every physician in the hospital has a 2-figure code number recorded on the next two columns, that every disease treated has a code number of 4 figures on the next four columns, and that one column records by one of 10 figures whether a patient is male or female, married or single, etc.

The operator will first set the machine to trap all cards with the code number for diabetes; he will then pass these cards through the machine again to trap those which record the code number of the physician; these again to trap those which record the code number of all women patients who are also married. Each of the remaining cards carries a code number corresponding to the names

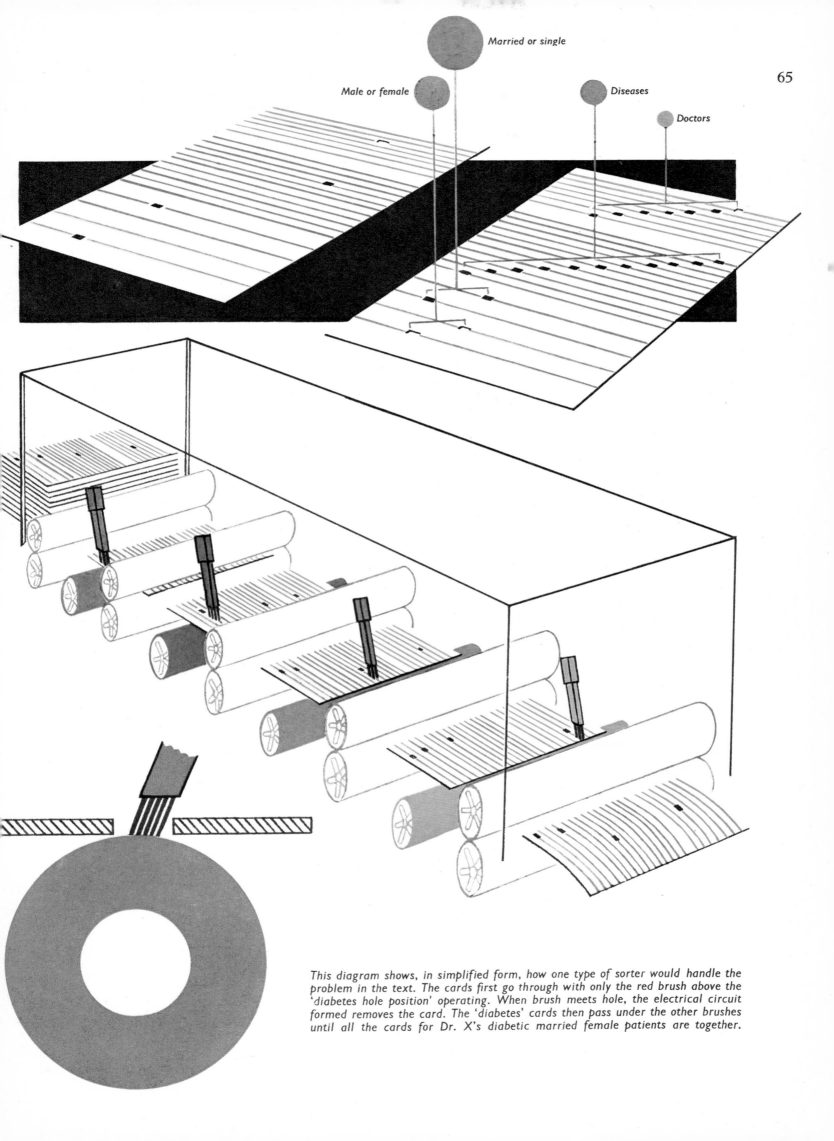

This diagram shows, in simplified form, how one type of sorter would handle the problem in the text. The cards first go through with only the red brush above the 'diabetes hole position' operating. When brush meets hole, the electrical circuit formed removes the card. The 'diabetes' cards then pass under the other brushes until all the cards for Dr. X's diabetic married female patients are together.

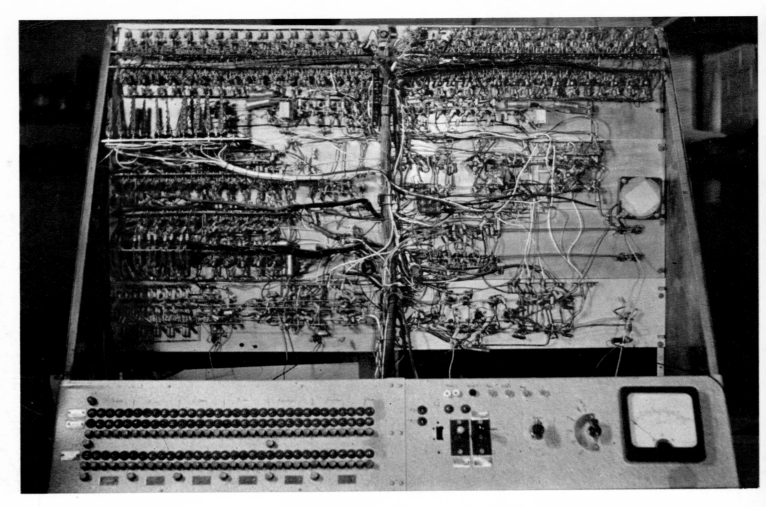

Electronic machine used for translation. Its magnetic storage drum (below) can take a grammar and a dictionary of up to 10,000 words.

Department of Numerical Automation, Birkbeck College

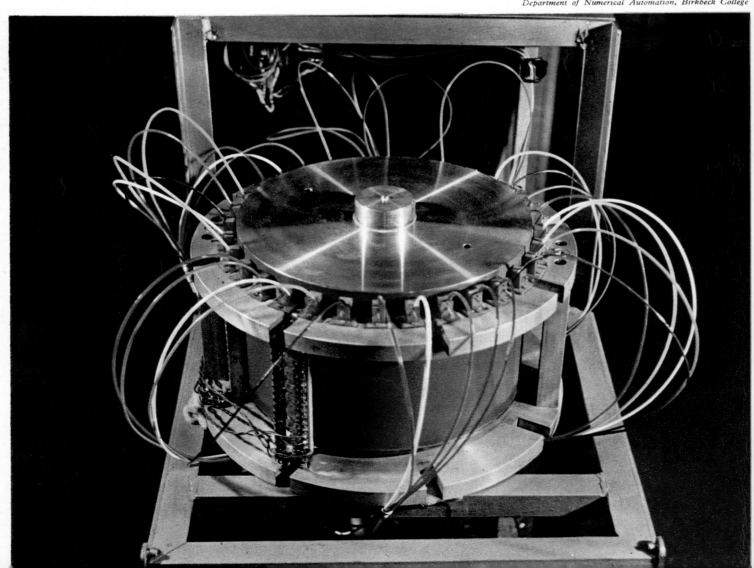

of a person of the category required. If the machine has a simple tabulating device, it will type off on tape a list of these code numbers. The operator can translate it; but it is also possible to construct a machine which can type the name itself from numbers of a code suitably designed for recording alphabetic spelling.

We have seen how a sorting machine can answer in our own language a question put to it in the only language it understands. If every patient of our parable had a nickname, it would be comparatively easy to make a machine which would record the nickname corresponding to a particular code number, and hence to a particular proper name. So making a machine to translate simple phrases from one language to another merely requires the construction of a dictionary giving a different code number for each corresponding pair of words, such as *égalité* (French)=2014=*equality* (English).

But it would be wrong to imagine that this is a model of the way in which engineers are actually trying to design a translation machine. We have now far more efficient and far less bulky ways of storing information than the use of punch cards. For instance, the magnetic tape. Also we have learned to rely more and more on a type of coding which does not employ the number-signs of our ordinary decimal system. None the less, the use of mechanical sorters and tabulators pointed the way to the construction of machines which can now perform more complicated acts than the translation of simple phrases.

Modern coding goes back to the telegraph principle of using two signs only. There is nothing sacred about the decimal system of ten signs (0, 1, 2, 3, 4, 5, 6, 7, 8, 9) except the fact that 0 placed at the end of a sequence always stands for multiplying by the same fixed base, which is 10 in our Hindu-Arabic signs. The base might equally well be 2 as we write it. There would then be no need for more than two signs, 1 and 0. Thus one would be written as 1, two as 10, three as 11, four as 100, five as 101, six as 110, seven as 111, eight as 1000, and so on.

Arithmetic would also be very simple, though wasteful of space when written down. All the rules needed would be $1 \times 0 = 0$; $0+1=1=1+0$; $1 \times 1 = 1$ and $1+1=10=1 \times 10$. Machines which record numbers swiftly and smoothly can calculate more simply this way and translate back into our way of writing numbers without worrying about space. They can also do something which is very important when translating from one language to another. When we set them to divide, they can jump several steps, as we do, by guesswork.

To interpret a long and difficult statement from one language to another involves a lot of such guesswork. To make a machine which can do this will certainly take a long time, for many reasons. Because so many words have many different meanings, the human translator has to rely on what he or she can gather from the intention of the writer before picking the right equivalent. Another difficulty is that the order in which people use words of different languages is different, and meaning depends much on the order of words. For instance, a blood red goat against an apple green hill does not mean the same as an apple green goat against a blood red hill.

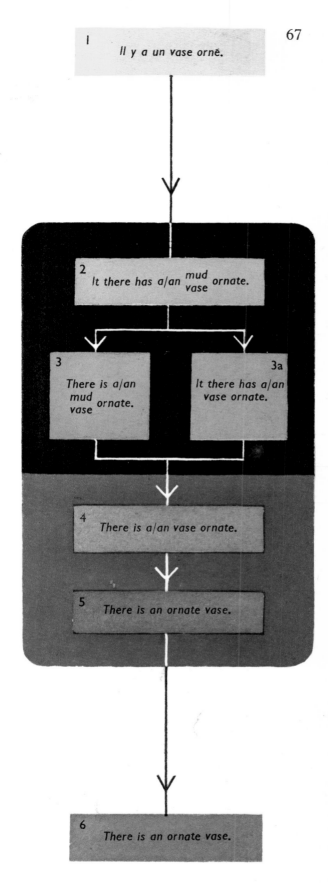

A. D. Booth
1. Sentence in first language on paper tape. 2. Word by word changed into second language. 3. Translation of idioms. 3a. Choice of correct word made where more than one is possible. 4. Sentence as corrected in 3 and 3a. 5. Words in correct order, phonetic ambiguity removed. 6. Sentence on tape now correct in second language. (Dictionary is shown black, grammar grey.)

If the attempt to make a machine do the job of the interpreter fails, there is a far simpler solution of how to make one human voice intelligible to all people living in the Radio Age. If all people would agree to adopt one and the same second language, children of the next generation would all be bilingual, like the Welsh and the Swiss. Like them, they would all have a second language in which to communicate with those who use other languages in the home; and this second language would be a language which all people everywhere would be able to speak and to understand.

So far, there have been three obstacles to getting such an agreement. One is that natural languages are far more difficult to learn than any of some four hundred concocted languages, such as Esperanto, devised to make the difficulty of learning as little as need be. A second is the fear that the use of any natural language, such as English, would give those who speak it in the home an unfair advantage. A third is that the choice of any made-up language not as yet widely used like English, in which so many translations are already available, would make the production of the first books on highly specialised subjects very expensive.

So our story ends here. There is now only one barrier to the power of mankind for communication instantaneously everywhere on our planet. A little good sense and goodwill could break through this sound barrier, if only more people keenly realised the need. The fact that governments will pay millions of dollars to make machines in the hope of making a break-through gives us hope that the next generation will do so, one way or another.

Kunsthistorisches Museum, Vienna – Farbwerke Hoechst AG.

Wherever the Old Testament is read, the story of the Tower of Babel is known. From early times people have recognised that the language barrier is one of the greatest handicaps to the progress of mankind. Now we have a chance to break through this barrier.

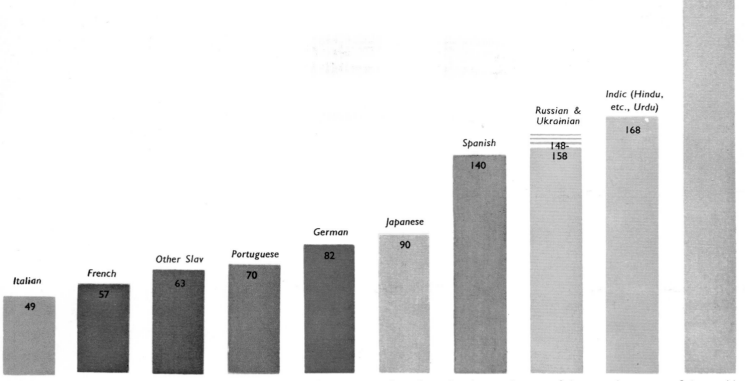

Based on figures from UNESCO, this diagram shows, in millions, the number of people who speak some of the main languages of the world.

Index

Advertisement	32, 33
Africa	8
African languages	57
Alexandria	20, 26
alphabet	12, 16, 17, 18, 19
arithmetic books	21, 24, 25
artists	23, 26, 27, 40, 41
Baird, John Logie	54
Behistun	14, 15
Bell, Alexander Graham	50
Bible	21, 25, 57
books	20, 22, 23, 26, 27, 68
Burma	39
calendar	8, 9
camera obscura	41, 42
camera, photographic	44
cave paintings	7, 8, 9, 10
Caxton	23
celluloid	45
C.G.S. system	61
Chaucer	21
China	12, 13, 23, 28, 36, 37, 39, 62
Chinese script	12, 13, 36
cigarette cards	33
cinema	45, 46, 47, 54
clay tablets	10, 11
Clerk-Maxwell	42
colour-printing	32, 33, 34, 35
computers	64
Crete	14, 15, 17
Cyprus	14, 17
Daguerre	42
Danish	56
Dante	21
dialects	56, 57
Dutch	56
Edison, Thomas Alva	45, 51
Egypt	10, 11, 12, 17
English	15, 18, 57, 62, 63, 69
Esperanto	68
France	21, 60, 61
Franklin, Benjamin	34
French	15, 18, 56, 62, 69
German	56, 62, 69
Germany	21, 61
Ghana	38, 39
Greece	61
Greek	15, 18, 19, 27, 56, 59, 61
Greeks	17, 18, 19, 20, 27
gunpowder	25, 28
Gutenberg	23
Hawaii	14
Helmholtz	42
Huss, John	25
Icelandic	18
illiteracy	20, 28, 30, 36, 38, 39
India	39, 62
Indo-China	39
Indo-European family	57
Italy	21, 40
Italian	56, 69
Japan	62
Japanese	13, 14, 69
Japanese script	13, 14
Kemal Ataturk	37
Kircher, Athanasius	41
language barrier	46, 54, 58, 63, 68
language families	19, 56, 57
Latin	19, 21, 56, 58, 59, 61
Laubach, Dr.	38
Lavoisier	58, 60
Linnaeus	58
Linotype machine	31, 32
Luther, Martin	28
magic lantern	41, 45
magnetic tape	11, 67
maps	26, 27
Marconi	53
master printers	27
Mayas	9, 11
medicine	27, 38
Mesopotamia	10, 11, 12
Middle East	10, 14
mining	25
money	10
monks	22
Monotype machine	32
Morse code	48
Moslem universities	20, 23
movable type	22, 23, 28
nautical almanacs	25
navigation manuals	21
Netherlands	61
newspapers	31, 32
Newton, Isaac	34, 42
Niepce	42
Nipkov	54
Norwegian	56
Ogham script	48, 49
Old English	56
optics	25, 40, 44
paper	22, 31
papyrus	11, 17, 19
parchment	22, 31
persistence of vision	44, 54
perspective	40
Petrarch	21, 23
Philippine Islands	14, 38
Phoenicians	16, 17
phonograph	46, 51, 64
photo-electric cell	54
photography	42, 44
picture-writing	11, 28
playing cards	22
Polynesia	38
Portuguese	69
printer's ink	23, 35
printing press	24, 25, 26, 27, 28, 31
radio	53
railways	48
Roumanian	18, 56
Russian	15, 18, 69
schools	21, 30
seals	10, 23
Semitic languages	17, 57
sign-writing	12
sorters and tabulators	64, 67
sound-shifts	18, 19, 56
Spain	20
Spanish	56, 69
spectacle-making	25, 40
spectrograph	34
stereotype	31
Sunday schools	30
Swedish	19, 56
syllable-writing	14, 15, 38, 39
Talbot, W. H. Fox	42
tape recorder	64
telegraph	48, 50, 67
telephone	50
television	54, 55
theatre	19
translation	62, 67
translation machine	63, 64, 66, 67
Turkey	37
UNESCO	38, 39
universal language	68
U.S.A.	63
U.S.S.R.	63
Wedgwood, Thomas	42
Welsh	63
windows	21, 40
wireless telegraphy	52, 53
wood-block printing	22, 23
word order	67
writing materials	17
writing tools	17
Wycliffe	21, 23, 25

On the endpapers the photograph of part of the stele of Tjetji is by R. L. Jarmain; it is reproduced by courtesy of the British Museum.